Biocontrol Agents of Phytonematodes

Biocontrol Agents of Phytonematodes

Editor

Rajan Jadhav

scitus
academics

Biocontrol Agents of Phytonematodes

Edited by **Rajan Jadhav**

ISBN: 978-1-68117-047-3

Library of Congress Control Number: 2015935442

SCITUS Academics LLC,
www.scitusacademics.com
Box No. 4766, 616 Corporate Way,
Suite 2, Valley Cottage,
NY 10989

Printed in United States of America on Acid Free Paper ∞

Contents

Preface

Nematophagous fungi are an important, fascinating group of soil microorganisms that can suppress the populations of plant-parasitic nematodes. They have been studied over several decades and the beautiful micrographs of these fungi trapping nematodes have marveled scientists and students alike. These remarkable fungi are also important potential biocontrol agents of nematodes diseases of plants and animals, they produce a swathe of natural chemicals and are model organisms for gene functional studies. The researcher has been working on many aspects of nematode-trapping fungi for more than 20 years. These studies have resulted in a wealth of information and it is now considered timely to bring all this information together into a book so that everyone who is interested can read about the amazing fungi. Previous studies based on traditional techniques have revealed much about nematophagous fungi. Modern techniques however, can now help to elucidate the molecular mechanisms underlying infection of nematodes by the nematophagous fungi, and help us to understand virulent factors, the role of proteases, chitinases and small chemical molecules, and the regulation of trap formation.

Editor

Chapter 1

Phylogenetic Analysis of Entomoparasitic Nematodes, Potential Control Agents of Flea Populations in Natural Foci of Plague

E. I. Koshel[1] V. V. Aleshin[2, 3, 4] G. A. Eroshenko[1], and V. V. Kutyrev[1]

[1]Russian Research Anti-Plague Institute "Microbe", Saratov 410005, Russia

[2]Belozersky Institute of Physical-Chemical Biology, Lomonosov Moscow State University, Moscow 119991, Russia

[3]Institute for Information Transmission Problems, Russian Academy of Sciences, Moscow 127994, Russia

[4]National Research Institute of Physiology, Biochemistry, and Nutrition of Farm Animals, Russian Academy of Agricultural Sciences, Kaluga Region, Borovsk 249013, Russia

ABSTRACT

Entomoparasitic nematodes are natural control agents for many insect pests, including fleas that transmitYersinia pestis, a causative agent of plague, in the natural foci of this extremely dangerous zoonosis. We examined the flea samples from the Volga-Ural natural focus of plague for their infestation with nematodes. Among the six flea species feeding on different rodent hosts (Citellus pygmaeus, Microtus socialis, and Allactaga major), the rate of infestation varied from 0 to 21%. The propagation rate of parasitic nematodes in the haemocoel of infected fleas was very high; in some cases, we observed up to 1,000 juveniles per flea specimen. Our study of morphology, life cycle, and rDNA sequences of these parasites revealed that they belong to three distinct species differing in the host specificity. On SSU and LSU rRNA phylogenies, these species representing three genera (Rubzovinema, Psyllotylenchus, and Spilotylenchus), constitute a monophyletic group close to Allantonema and Parasitylenchus, the type genera of the families Allantonematidae and Parasitylenchidae (Nematoda: Tylenchida). We discuss the SSU-ITS1-5.8S-LSU rDNA phylogeny of the Tylenchida with a special emphasis on the suborder Hexatylina.

INTRODUCTION

More than 150 species of fleas feeding on different mammalian hosts, primarily rodents, are vectors of the bacterium Yersinia pestis, a causative agent of plague [1, 2]. In natural foci of plague, the dynamics of flea populations are among the main factors controlling the incidence of epizootics that pose a threat to humans inhabiting the areas [3–5]. Entomoparasitic nematodes of the order Tylenchida are known to control populations of various insect hosts [6–9]. The rate of tylenchid infestation in fleas reaches 50–60% in some cases [10, 11], when the nematodes cause castration and early death of the flea hosts [9, 12, 13].

Despite high importance of the Tylenchida as a nematode order harboring entomoparasites and notorious crop pests, their reliable phylogeny is still a challenge. Tylenchid nematodes differ widely in life cycle, parasitic strategies, and the host range that spans plants, fungi, and invertebrates. Phylogenies obtained from SSU and partial LSU rDNA data often disagree with classifications based on morphology and life cycle [14–21]. Phylogenetic resolution inside the order is far from being clear, which in many respects results from the insufficiency of data available to adequately describe its diversity. As for tylenchid parasites of fleas, only 31 species are described to date [9, 22–31], with no molecular vouchering. Here we present a study of parasitic nematodes isolated from fleas sampled from different rodent hosts in a natural focus of plague.

MATERIALS AND METHODS

Collection of Samples

Samples were collected in 2012 (spring and autumn) and 2013 (spring) in the Volga-Ural natural focus of plague (Figure 1). The sampled rodents included sousliks (Citellus pygmaeus), mouse-like rodents (Microtus socialis and Apodemus uralensis), and jerboas (Allactaga major). Three flea species (Citellophilus tesquorum,Neopsylla setosa, and Frontopsylla semura) were sampled on sousliks; two species (Amphipsylla rossica andCtenophthalmus secundus) were on M. socialis voles; and one species (Mesopsylla hebes) was on jerboas. Fleas were examined for nematode infestation (Table 1). Examination and dissection of fleas were carried out using the dissecting microscope MBS-2 (LOMO, Russia). A half of parasitic nematodes sampled from each flea was preserved for subsequent DNA extraction, and another half was used for morphological analysis. Live fleas infected with nematodes were placed in glass flasks with river sand to obtain free-living forms. Insects were kept in a KBF 720 (E5.2) climate chamber (Binder, Germany) at 26°C and 80% humidity.

Table 1: Number of fleas studied and the percentage of fleas infected with nematodes

Time of sampling	Host rodent species	Flea species	Number of collected fleas	Number of infected fleas	Percentage of infected fleas
April 2012	Citellus pygmaeus	Citellophilus tesquorum	41	7	17.1%
		Neopsylla setosa	73	5	6.8%
		Frontopsylla semura	54	7	13%
October 2012	Microtus socialis	Amphipsylla rossica	135	9	6.7%
		Ctenophthalmus secundus	88	1	1.1%
April 2013	Citellus pygmaeus	Citellophilus tesquorum	34	0	0
		Neopsylla setosa	271	22	8.1%
		Frontopsylla semura	19	4	21%
	Microtus socialis and Apodemus uralensis	Amphipsylla rossica	6	0	0
		Ctenophthalmus secundus	52	0	0
	Allactaga major	Mesopsylla hebes	34	2	5.9%

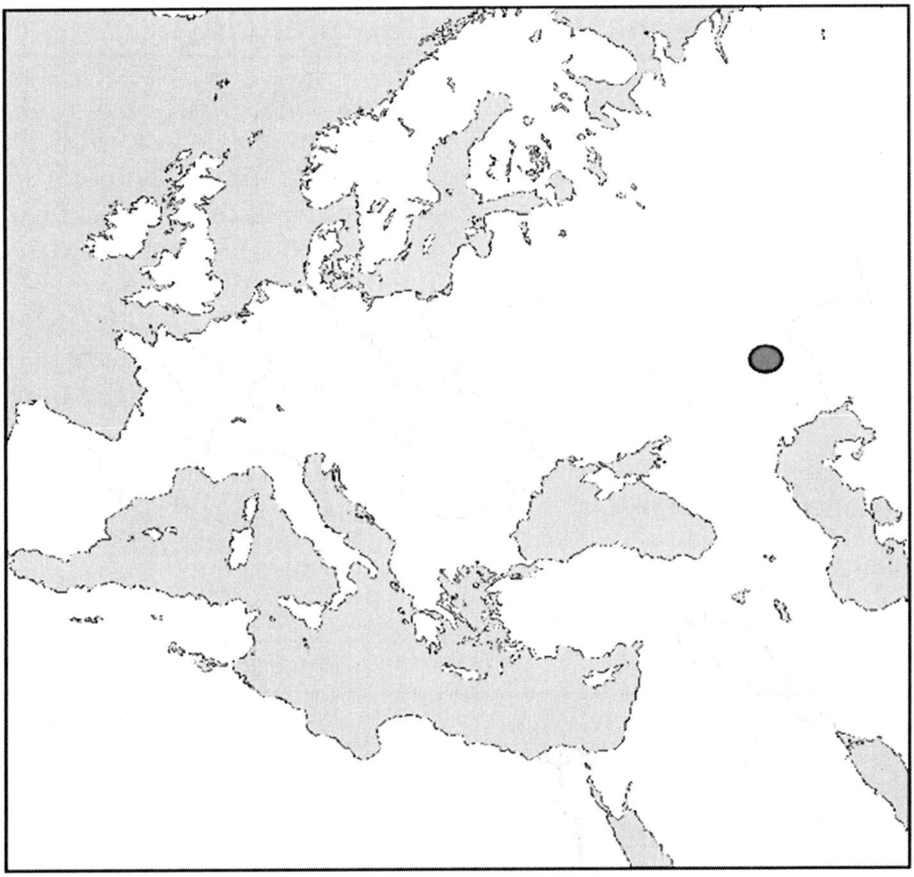

Figure 1: The sampling region on the map of Europe.

Morphological Analysis

Fixation and clarification of nematode preparations were performed using standard techniques described by De Grisse [32]. Material was mounted on slides in a drop of glycerin, bound by a paraffin circlet (http://pest.cabweb.org). Color staining of preparations was not performed. Morphometric analysis was conducted using the light microscope "Leica DM 1000" (Leica, Germany) with an eyepiece micrometer. Pictures of nematodes were taken with the microscope "DFC 425" (Leica, Germany). Published data on morphometrics [23, 25, 26] were used for comparison.

DNA Extraction, PCR, and Sequencing

DNA samples were extracted with a Diatom DNA Prep (IsoGen Lab, Russia). rDNA fragments were amplified using an Encyclo PCR kit (Evrogen, Russia) and primers given in Table 2. The amplified rDNA fragments were sequenced using an Applied Biosystems 3500xL DNA analyzer. Sequence reads were assembled with the CAP contig assembly program [33] and proofread with the BioEdit software [34]. For three isolates, almost complete sequences of 18S and 28S rRNA and complete sequences of 5.8 rRNA, internal transcribed spacers ITS1 and ITS2 were assembled. The sequences were submitted to GenBank under accession nos. KF155281–KF155283. For the rest of isolates, partial (750–800bp) sequences of 18S and 28S rRNA genes were submitted to GenBank under accession nos. KF373731–KF373740.

Table 2: Nucleotide sequences of primers used in this study

Primer	Sequence	Orientation	References
Nik22	tmycygrttgatyctgyc	F	This study
A	gtatctggttgatcctgccagt	F	[35]
Q5nemCh	gccgcgaayggctcattayaac	F	This study
G18SU	gcttgtctcaaagattaagcc	F	[36]
Ves18-d9	gtcgtaacaaggtatccgtaggtgaac	F	This study
R18Tyl1	ggtccaagaatttcacctctc	R	[36]
B	gtaggtgaacctgcagaaggatca	R	[35]
Q39nem	gaaaccttgttacgacttttrcbygg	R	This study
58d1	rcatcgatgaagaacgywg	F	[37]
58r nem	gcwgcgttcttcatcgacyc	R	This study
28d3	gtcttgaaacacggaccaagg	F	[37]
28d6	ggtyagtcgrtcctrag	F	[37]
D2A	acaagtaccgtgagggaaagttg	F	[38]
28r4	gctatcctgagggaaacttcgg	R	[37]

28r2nem	cggtacttgttcgctatcg	R	This study
28r7	agccaatccttwtcccgaagttac	R	[37]
28r12	ttctgacttagaggcgttcag	R	[37]
D3B	tcggaaggaaccagctacta	R	[38]

Phylogenetic Analysis

The newly obtained rDNA sequences of tylenchid parasites of fleas were aligned with a selected set of other tylenchid sequences obtained from the GenBank. The main selection criterion was to sample representatives of all clades that occur in published SSU and LSU rDNA phylogenies of the Tylenchida [16–21, 39]. Apart from the D2-D3 LSU rDNA expansion segment commonly used in previous studies, we included all LSU rDNA sequence data available for the Tylenchida, with the exception of Basiria sp. SAN-2005 (accession nos. DQ145619, DQ145667) that in our preliminary analyses (data not shown) demonstrated a disputable affinity to the Tylenchida. For the species Anguina tritici, Globodera pallida, Heterodera glycines, Pratylenchus vulnus, andRadopholus similes the nearly complete rDNA sequences were assembled with appropriate cDNA fragments identified with BLAST [40]. Partial LSU rDNA sequence of Ditylenchus dipsaci was combined with the soil environmental clone NTS_28S_061A_2_b4 (accession no. KC558346), as the clone sequence appeared to represent a close tylenchid relative of D. dipsaci. Chimeric sequences were also created in some cases when closely related partial rDNA sequences were found in the database. All sequences and their accession numbers are listed in Table 3. Cephalobidae and Chambersiellidae were chosen as the out group. Alignments were constructed with the MUSCLE program [41] and refined manually using the MEGA 5.0 software package [42]. Three alignments were generated: (1) SSU rDNA, (2) D3 region of LSU rDNA, and (3) concatenated rDNA data including SSU, LSU, 5.8S rDNA, and highly conserved regions of ITS1. After discarding ambiguously aligned positions, the alignments length was 1,723, 592, and 4,930 positions, respectively. Bayesian reconstruction of phylogeny was done with the PhyloBayes software, version 3.2 [43] under the GTR + CAT + DP model [44]. Eight independent runs were performed with 4,000,000 cycles each; the first 3,000,000 cycles were discarded. A consensus tree with Bayesian posterior probabilities was constructed for the remained tree sample. Bayesian reconstruction was also performed using the MrBayes software [45] under the GTR + G8 + I model [46] in two independent runs, each with four Markov chains. The chains were run for 5,000,000 generations, with trees sampling every 1,000th generation. The consensus posterior probabilities were calculated after discarding the first 3,000,000 generations. Partitioning "by genes" was used for the concatenated alignment with all parameters unlinked, except for the topology and branch lengths. In addition, node support was estimated with maximum likelihood bootstrap as implemented in the RAxML software, version 7.2.6 [47], under the GTR + G + I model with 1,000 bootstrap replicates. Alternative topologies were tested using the approximately unbiased (AU) [48] and Kishino and Hasegawa [49] tests implemented in the CONSEL software [50] and the expected likelihood weight test [51] implemented in the TREE-PUZZLE software [52]. TREEVIEW [53] was used as the tree viewer and editor, and site-wise log-likelihoods were computed with TREE-PUZZLE under the GTR + G8 + I model with substitution matrix parameters estimated by MrBayes.

Table 3: List of OTUs and accession numbers of sequences

Name	18S rRNA	ITS1-5.8S rRNA	28S rRNA	%, SSU-ITS1-5.8S-LSU/D3	Reference	Family by [8]
Chambersiellidae*						
Fescia grossa	KC242218	—	DQ145636 DQ145684	87.1/—	[54] [55]	Chambersiellidae
Geraldius sp. SAN-2010a	—	—	GU062821	17.8/—	[56]	Chambersiellidae
Cephalobidae						
Acrobeloid s maximus	EU196016	JX026706	EU195987	94.8/—	[57] [58] [57]	Cephalobidae
Cephalobus cubaensis	AF202161	AF202161	EU253570	89.8/—	[59] [57]	
Panagrolobus sp. SN-2010	—	—	HM439771	51.9/—	[60]	
Cephalobidae Gen. sp. MHMH-2008	FJ040406	—	—		Holterman et al., 2008, unpublished.	
Zeldia punctata	—	DQ146426	EU195988	96.6/—	[61] [57]	
Zeldia sp.	AY284675	—	—			
Aphelenchidae						

Aphelenchus avenae	JQ348399	AF119048	—	96.9/—	[62] [63]	Aphelenchidae
Aphelenchus s .	—	—	DQ145664 DQ145714		[55]	
Paraphelenchus acontioides	—	—	HQ218322	45.5/—	[64]	
Paraphelenchus sp.	AY284642	—	—		[18]	
Hexatylina + "Anguinata (art)": Iotonchioidae						
Allantonema mirable	—	—	JX291132	10.6/85.8	[39]	Allantonematidae
Bradynema listronoti	DQ915805		DQ915804	45.6/96.8	[65]	
Bradynema rigidum			DQ328730	10.4/86.3	[20]	
Contortylenchus sp.	—	—	DQ328731	—/85.4	[20]	
Deladenus duru	JQ957898	—	—	34.0/—	[66]	Neotylenchidae
Deladenus proximus	JF304744	JF304744	—	35.2/—	[67]	
Deladenus siricidicola isolate 354	AY633447		AY633444	45.8/98.1	[68]	
Deladenus siricidicola isolate 466	FJ004890	FJ004890	—	41.7/—	[69]	
Deladenus siricidic la isolate 1093	FJ004889	FJ004889	—	42.0/—	[69]	
Fergusobia camaldulensae	AY589294	—	AY589346	45.7/98.0	[68]	
Fergusobia sp. 444	EF011667	—	EF011675	45.7/97.3	[68]	
Fergusobia sp. SBG	FJ393270	—	FJ386996	45.7/98.3	[70]	
cf. Gymnotylenchus sp. TSH-2005	AY912040	—	—	12.9/—	Powers et al., unpublished.	

Howardula aoronymphium	AY589304	AY589304	AY589395	49.7/96.1	[68]	Allantonematidae
Howardula dominicki	AF519234	AF519234	—	37.4/—	[71]	
Howardula neocosmis	AF519226	AF519226	—	38.2/—	[71]	
Howardula phyllotretae	JX291137	—	DQ328728	41.9/86.1	[39] [20]	
Howardula sp. CD353	—	—	JX291131	—/93.9	[39]	
Howardula sp. SP-A	AF519232	AF519232	—	37.7/—	[71]	
Howardula sp. SP-F	AF519222	AF519222	—	38.2/—	[71]	
Howardula sp. SP-MA	AF519233	AF519233	—	38.1/—	[71]	
Howardula sp. SP-PS	AF519231	AF519231	—	38.1/—	[71]	
Parasitylenchus bifurcatus	KC875397	—		44.0/85.3	[72]	
Parasitylenchus sp.	—	—	DQ328729		[20]	
Psyllotylenchus sp. ex Frontopsylla semura	KF373734	—	KF373739	27.1/93.7	This study	Parasitylenchidae
Psyllotylenchus sp. ex Neopsylla setosa	KF373733	—	KF373738	27.1/9 .7	This study	
Rubzovinema sp. ex Amphipsylla rossica	KF155281	KF155281	KF155281	90.0/100.0	This study	Neotylenchidae
Rubzovinema sp. ex Ctenophthalmus cecundus	KF155282	KF155282	KF155282	89.8/100.0	This study	
Rubzovinema sp. ex Citellophilus tesquorum	KF155283	KF155283	KF155283	93.2/100.0	This study	
Rubzovinema sp. ex Frontopsylla semura	KF373732	—	KF373737	27.1/93.7	This study	
Rubzovinema sp. ex Neopsylla setosa	KF373731	—	KF373736	27.1/93.7	This study	
Skarbilovinema laumondi	—	—	JX291136	10.9/91.0	[39]	Iotonchioidea
Skarbilovinema lyoni	JX291138	—	DQ328733	41.8/86.3	[39] [20]	

Spilotylenchus sp. ex Mesopsylla hebes	KF373735	—	KF373740	27.1/93.4	This study	Parasitylenchidae
cf. Sychnotylenchus sp. CSP1-09	DQ080531	—	—	12.9/—	Powers et al., unpublished.	Sychnotylenchidae
Wachekitylenchus bovieni	—	—	DQ328732	—/85.9	[20]	Parasitylenchidae
Unidentified Allantonematidae HaMW	JQ941710	—	—	18.5/—	Rhule, unpublished.	Allantonematidae
Unidentified Allantonematidae NK2011_2	AB663183	—	—	12.0/—	[73]	
Unidentified Allantonematidae NK2011_3	AB663184	—	—	12.0/—	[73]	
Unidentified nematode 804U-025	EU880149	—	—	12.0/—	[74]	
Unidentified nematode CD289	—	—	JX291133	—/84.1	[39]	
Unidentified nematode RGD591T12	AB455970	—	—	12.0/—	[73]	
Unidentified ematode WY2009_BAR-1	—	—	FJ661075	—/96.3	[75]	
Unidentified parasite exChrysobothris affinis	—	—	DQ202658	—/51.0	Hunt et al., unpublished.	
Hexatylina + "Anguinata (part)": Sphaerularioidea						
Deladenus sp. PDL-2005	AJ966481	—	—	35.0/—	[16]	Neotylenchid e
cf. Helionema sp. MHMH-2008	EU669913	—	—	34.0/—	[19]	Parasitylenchidae (genera dubia in Hexatylina)
cf. Hexatylus sp. Westplace	AY912050	—	—	12.9/—	Powers et al., unpublished.	Neotylenchidae
Nothotylenchus acris	AY593914	—	—	34.0/—	[76]	Anguinidae

Sphaerularia bombi	AB250212	—	DQ328726	56.7/100.0	Takahashi, unpublished.20]	Sphaerulariidae
Sphaerularia vespae	AB300595	AB300595	AB300596	54.7/100.0	[77]	
Unidentified nematode 801L-022	EU880129	—	—	12.1/—	[74]	
Anguinata						
Anguina tritici	AY593913	JF826515	HO058555 DQ328723	57.6/92.9	Holterman et al., unpublished. Rao and Rao, unpublished. Rao et al., unpublished. [20]	Anguinidae
Ditylenchus adasi	EU669909	—	—	34.6/—	[19]	
Ditylenchus angustus	AJ966483	—	—	3.6/—	[16]	
Ditylenchus destructor		JX162205		50.0/99.5	[78]	
Ditylenchus dipsaci	AY593911	AY593911	JF327759	60.9/100.0	[76]Zhao 2011, unpublished.	
clone NTS_28S_061A_2_b4			KC558346		[79]	
Ditylenchus drepanocercus	JQ429768	JQ429774	JQ429772	48. /89.3	[80]	
Ditylenchus halictus	AY589297			52.8/97.3	[68]	
Ficotylus congestae	EU018049			45.6/97.5	[81]	
Halenchus fucicola	EU669912	—	—	34.6/—	[19]	
Pseudhalenchus minutus	AY284638			34.6/—	[19]	
Unidentified entomoparasitic nematode SAS-200 "Neotylenchus" sp.	—	—	DQ328725	—/85.6	[20]	
"Tylenchina": Tylenchidae						

Aglenchus agricola	FJ969113	—	—	46.0/—	van Megen et al., unpublished.	Tylenchidae
Aglenchus sp.	—	—	JQ004996		[82]	
Coslenchus costatus	AY284581	—	—	45.5/—	[18]	
Coslenchus sp.		—	JQ005007		[82]	
Filenchus annulatus	JQ814880	—	JQ005017	46.4/—	[82]	
Tylenchus davainei	AY284588	—	—	33.9/—	[18]	
"Tylenchina": Tylodoridae						
Eutylenchus excretorius	EU915487	EU915500	EU915490	35.8/—	[83]	Atylenchidae
Cephalenchus hexalineatus	AY284594	—	—	44.1/—	[18]	Tylodoridae
"Tylenchina": Boleodoridae						
Basiria gracilis	EU130839	—	DQ328717	44.6/—	[84] [20]	Tylenchidae
Basiria sp. 3 TJP-2012	—	—	JQ004998	12.0/—	[82]	
Boleodorus thylactus	AY993976	—	—	46.7/—	[16]	
Boleodorus sp.	—	—	JQ005001		[18]	
Neopsilenchus magnidens	AY284585	—	—	45.6/—	[18]	
Neopsilenchus sp. 3 TJP-2012	—	—	JQ005020		[82]	
Neopsilenchus sp. 1 TJP-2012	—	—	JQ005018	11.9/—	[82]	
"Hoplolaimina": Merliniidae						
Nagelus leptus	—	—	DQ32 715	45.2/—	[20]	Telotylenchidae
Nagelus obscurus	EU306350	—	—		[17]	
Pratylenchoides ritteri	AJ966497	—	JX261964	48.7/—	[16] [85]	Pratylenchidae

Psilenchus cf. hilarulus	AY284593	—	EU915489	44.1/—	[18] [83]	Psilenchidae
Scutylenchus quadrifer	AY284 99	—	—	41.5/—	[18]	Telotylenchidae
Scutylenchus sp.	—	JQ069956	—		[86]	
"Tylenchina": Ecphyadophoridae						
Ecphyadophora sp. JH-2004	AY593917	—	—	33.7/—	[76]	Ecphyadophoridae
"Ditylenchus" brevicauda	AY284635	—	—	33.9/—	[18]	Anguinidae
Malenchus andrass i	AY284587	—	—	32.3/—	[18]	Tylenchidae
Ottolenchus discrepans	AY284590	—	—	33.7/—	[18]	
Criconematina						
Hemicriconemoides gaddi	—	KC520471	KC520470	55.6/—	[87]	Criconematidae
Hemicriconemoides pseudobrachyurus	AY284622	—	—		[18]	
Hemicycliophora lutos	—	Q406237	GQ406240	53.2/—	[88]	Hemicycliophoridae
Hemicycliophora thienemanni	AY284628	—	—		[18]	
Meloidoderita kirjanovae	—	DQ768427	DQ768428	50.8/—	[89]	Sphaeronematidae
Sphaeronema alni	FJ969127	—	—		van Megen, unpublished.	
Meloidoderita sp.	GU 53916	GU253917	JQ771954	50.8/—	[90] Cudejkova and Cermak, unpublished.	

Tylenchulus semipenetrans	AJ966511	FJ588909	FJ969710	57.5/—	[16] [91] [92]	Tylenchulidae
"Hoplolaimina": Belonolaimidae						
Belonolaimus longicaudatus	AY633449	DQ67236	GQ896548	55.8 —	[68] [93] [94]	Belonolaimidae
Ibipora lolii	JQ771535	—	—	30.9/—	[95]	
"Hoplolaimina": Hoplolaimidae						
Carphodorus sp.	JQ771538	—	JQ771550	41.3/—	[95]	

Globodera pallida	EU855119	EU85511	BM415342 BM415248 CV577211 CV577977 V579301E U85511	93.6/—	Nowaczyk et al., unpublished. Opperman, unpublished [96].	Heteroderidae
Heterodera glycines	AF216579 BI704127. BI748392. CA940548 CB379240. CB379263 CB379850 CB380242 CB825296 CB825409 CB825970 CB935610 CK348871 CK348904 CK349175 CK352112	AF216579	AF133304 AF216579 BI704144 BI704144 BI749520 CA940190 CA9402 12 CA940243 CA940406 CA940424 CA940429 CA940589 CB238697 CB279977 CB299455 CB373844 CB373981 CB379125.	98.3/—	[97] [96]. [98] Yan and Davis, unpublished. [99] Ye et al., unpublished. Wei et al., unpublished.	

			CB379140 CB379219 CB379312 CB379439 CB379505 CB379696 CB379707 CB379996 CB380091 CB380241 CB824788 CB824878 CB825995 CB934877 CB934931 CB934950 CB934954 CK348525 CO036619 HM560850 JN684906			
Morulaimus sp.	JQ771540	—	—	31.5/—	[95]	Belonolaimidae

Radopholus s milis	AJ966502 AY912509 EF384224 EY190988 EY191076 EY191697 EY191883 EY192786 EY192788 3EY193123 EY193253 EY194340 EY194464 EY194646 EY195472 FJ040398	AY912509 EF384224	EU555409 EY189839 EY190550. EY190620 EY190961 EY191066 EY191073 EY191135 EY191160 EY191173 EY191237 EY192021 EY192028 EY192080 EY192091 EY192247 EY192381 EY192472 EY192501 EY192526 EY192892 EY192907 EY1930 05 EY193037. EY1932 9 EY193314 EY193798 EY193897 EY193971 EY194395 EY194454 EY194530 EY195146 EY195204 EY195406	97.5/—	[16] [100] Long et al., unpublished. [101] Holterman et al., unpublished. [102] [100] Zhao unpublished. [86]	Pratylenchidae

			EY195408 EY195580 EY195889 EY195943 GQ281471 JN091962 JQ782249			
Rotylenchulus reniformis	JX406356	FJ374686	HM13188 FJ906072	59.4/—	[103] Rahman et al., unpublished. [104]	Rotylenchulidae
"Hoplolaimina": Pratylenchidae						
Dolichodorus sp. WY-2006	DQ912918	—	—	33.9/—	[105]	Dolichodoridae
Hirschmanniella loofi	EU306353	EU620472	EU620469	51.6/—	[17] [106]	Pratylenchidae
Macrotrophurus arbusticola	AY284595	—		33.9/—	[18]	Telotylenchidae
Meloidogyne arenaria	U42342	U42342	U42342 AF023855 AF023856	99.2/—	Georgi and Abbott, unpublished.	Meloidogynidae
Meloidogyne artiellia	AF248477	AF248477	AF248477	99.2/—	[107]	
Nacobbus aberrans	AJ966494	DQ017473	U47557 9.0/—		[16] [108] [109]	Pratylenchidae

Pratylenchus vulnus	EU669955	JQ966892	BQ580554 CV198923 CV198995 CV199233. CV199349 CV199490 CV200136 CV200423 CV200464 CV200467 CV200471 CV200530 CV200687 CV200896 CV201004 CV201135 EL887566 EL887705 EL888035 EL888060 EL888174 EL888269 EL888739 EL888778 EL889241 EL889472 EL889797 EL889934 EL889934 EL889977 EL889994 EL890380 EL890701 JQ003993 JQ003994 JX047008	100.0/—	[19] [110] [96] [96] [111] Zhao, unpublished. [112]	

Tylenchorhynchus dubius	EU306352	—	DQ328707	53.2/—	[17] [20]	Telotylenchidae
Tylenchorhynchus zeae	—	EF519711	—		[113]	

*Clades of the tree, marked by boldface.

RESULTS

Infestation of Fleas with Nematodes

The infestation rate is shown in Table 1 (in total, 807 flea specimens were studied). Among the six flea species studied, the population size and the percentage of infected fleas varied depending on the season. Three flea species sampled on sousliks (Citellophilus tesquorum, Neopsylla setosa, and Frontopsylla semura) exhibited a stable population density. In the two species, N. setosa and F. semura, the infestation rate was moderate to high in the spring seasons of 2012 and 2013. In C. tesquorum, no infected fleas were detected in spring 2013, whereas in spring 2012 the fleas were highly infested (17.1%). The vole flea Amphipsylla rossica was abundant and moderately infested in autumn, whereas being less abundant in spring, which may explain the absence of infected fleas in the spring sample. Another vole flea, Ctenophthalmus secundus, exhibited a consistently high population density and low infestation rate in both spring and autumn samples.

Adult parasitic females and their progeny were found in the haemocoel of infected fleas. In the infected fleas C. tesquorum, A. rossica, C. secundus, and Mesopsylla hebes, only one generation of parasitic females was observed. Their amount in a flea specimen is determined by the number of free-living infective females that penetrate into the flea larva. We observed 1 to 2 or 1 to 4 adult parasitic females per flea specimen in spring and autumn, respectively. An additional parthenogenetic generation of parasitic females was found in some fleas ofN. setosa and F. semura, where up to 16 specimens per flea were observed. As in other entomoparasitic nematodes, the propagation rate depends on the host age. Thus, in young fleas up to 10 juveniles was found per flea specimen, whereas up to 1,000 juveniles of different stages were contained in some old fleas (Figure 2). After the 2nd molt the number of juveniles is maximal, and 3rd stage juveniles massively migrate to the rectal section of the flea intestine for exit to the environment. In some cases, the observed infestation level was so high that nematodes penetrated distal segments of the flea legs, from where they have no way to the environment.

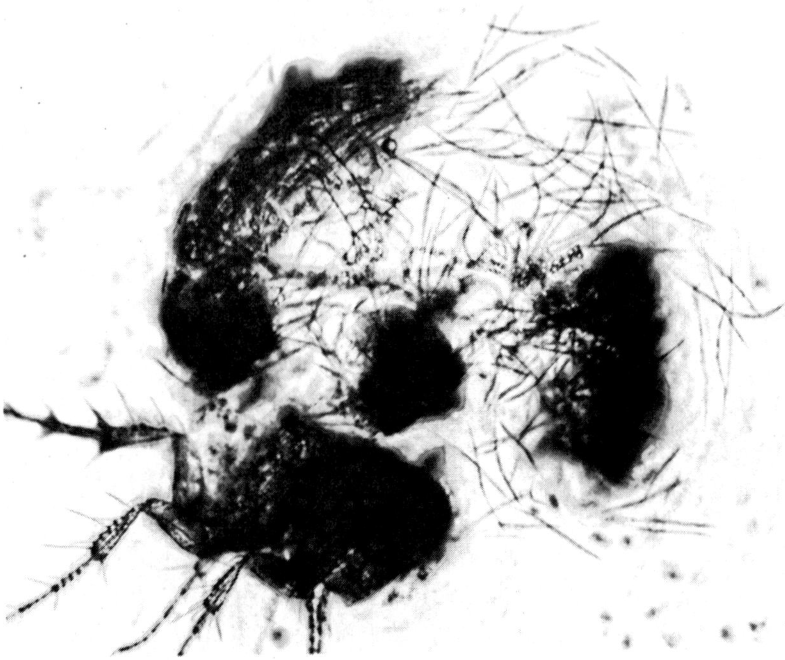

Figure 2: Numerous juveniles of Rubzovinema sp. extracted from the dissected body of aCitellophilus tesquorum flea.

Morphological Analysis of Entomoparasitic Stages in Nematode Isolates and Their Taxonomic Identification

Analysis of morphology of entomoparasitic stages suggests that the studied nematode isolates from three distinct groups. A single generation of parasitic females was observed in the first two groups and an additional parthenogenetic generation—in the third group. According to morphometric data on adult parasitic females (Tables 4–6), the first two groups belong to the genera Rubzovinema or Spilotylenchus and the third group to the genus Psyllotylenchus. Photographs of parasitic females of Rubzovinema sp., Spilotylenchus sp., andPsyllotylenchus sp. are depicted in Figure 3. Figure 4 shows their distribution among flea samples studied.

Table 4: Comparison of morphometrics in parasitic females of Rubzovinema sp. And Rubzovinema ceratophylla.

Character	Rubzovinema sp. (this study)	Rubzovinema ceratophylla [26]
N	29	27
L	1278,6 (840–1570)	1265,1 (810–1840)
D	120,8 (85–145)	137,3 (62–200)
A	11,19 (7,9–16,1)	9 51 (6,4–16,8)
C	65,4 (31,4–100)	44,10 (10–86,4)
V%	96,4 (93,1–97,9)	95,44 (92–98,9)
Total length of stylet (St)	18,5 (14–22)	19,5 (18–21)
Length of distal edge of stylet	7,2 (5–8,7)	—
Distance between anterior end and excretory pore (Ex)	20,7 (10–31)	—
Distance between anterior end and nerve ring	61,2 (50–74,5)	
Total length of tail (Cd)	21,9 (10–42)	26,35 (14–47,5)
Distance between vulva and tail end	46,1 (23–75)	—
Distance between vulva and anus (V–A)	26,9 (13–40)	—

All measurements are in m and in the form mean (range).

Table 5: Comparison of morphometrics of parasitic females in Spilotylenchus sp. andSpilotylenchus maisonabei

Characters	Spilotylenchus sp. (this study)	Spilotylenchus maisonabei[23]
N	2	6
L	1,600–1,840	1,244 (1,200–1,320)
D	155–160	125 (107–160)
A	10.3–11.5	10.3 (7.5–12)
C	167.3–177.8	84.4 (64.5–121)
V%	97.4–97.7	96.2 (95.8–96.5)

Total length of stylet (St)	9.5–9.8	9-10
Distance between anterior end and excretory pore	1.5–15.5	23.3 (20–28)
Distance between anterior end and nerve ring	—	52–54
Total length of tail (Cd)	9–11	15.4 (10–19)
Distance between vulva and tail end	41.5–43	47 (42–52)
Distance between vulva and anus (V–A)	32-33	—

All measurements are in m and in the form mean (range).

Table 6: Comparison of morphometrics of parasitic females in Psyllotylenchus sp. And Psyllotylenchus viviparous

Character	Psyllotylenchus sp. (this study)		Psyllotylenchus viviparous [25]	
	Gamogenetic	Parthenogenetic	Gamogenetic	Parthenogenetic
N	3	7	8	10
L	1,016.7 (900–1,100)	446 (420–500)	1,000 (840–1,480)	500 (360–840)
D	81.3 (79–84)	70 (60–80)	77 (62–115)	60 (54–100)
A	12.5 (11.1–13.3)	6.25 (5.6–7)	—	—
C	64.3 (60–68.2)	40.15 (37.1–43.5)	—	—
V%	95.1 (95–95.4)	93.3 (90–95.3)	—	—
Total length of stylet (St)	17.5 (17–18,5)	5.25 (4–6)	17 (15–20)	7 (5–8)

Length of the distal edge of stylet	8.6 (8-9)	—	—	—
Distance between anterior end and excretory pore	26.5 (25–31.5)	17.5 (15–19.5)	23 (13–33)	22 (14–46)
Distance between anterior end and nerve ring	—	51.7 (50–55)	—	—
Total length of tail (Cd)	15.8 (15–17)	11.1 (10.5–11.5)	25 (17–35)	9 (1–17)
Distance between vulva and tail end	48 (45–51)	30.5 (19.7–55)	56 (37–71)	52 (40–104)
Distance between vulva and anus (V–A)	30.8 (29–31.5)	13.5 (11.7–21.6)	—	—

All measurements are in m and in the form mean (range).

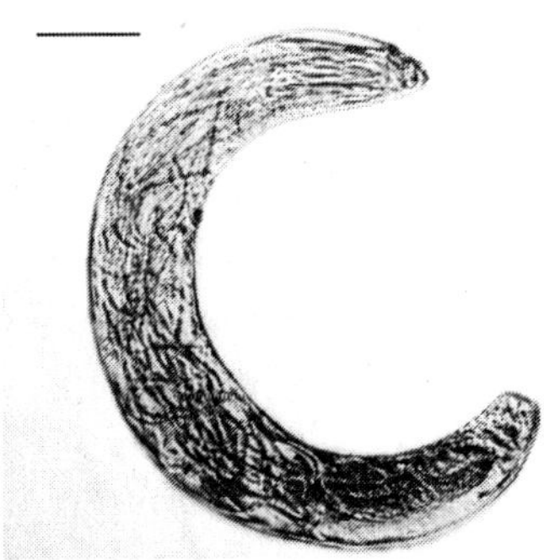

Figure 3: Parasitic females of the studied nematode species. (a) Rubzovinema sp., heterogeneous female; (b) Spilotylenchus sp., heterogeneous female;

(c) Psillotylenchus sp., heterogeneous female of the first generation; (d) (c): Psillotylenchus sp., parthenogenetic female of the second generation. Scale bar—200 μm.

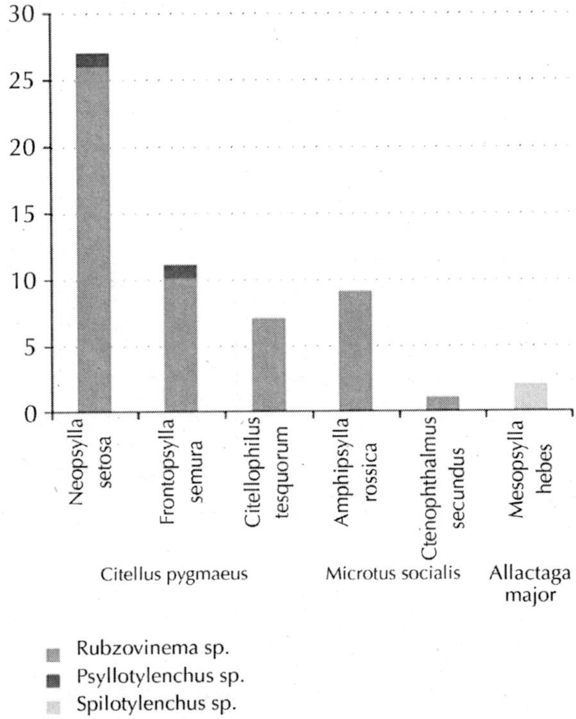

Figure 4: Distribution of the studied nematode species among the flea species studied, whose rodent hosts are given below. The vertical axis shows the numbers of infected fleas.

According to morphometric evidence, parasitic females and juveniles of the genera Rubzovinema andSpilotylenchus are very similar. However, in the first two groups of isolates we found characters bearing discriminative and identificational value. In particular, the oesophageal glands in juveniles III of the first group are poorly developed. This is a distinctive feature of the genus Rubzovinema, where males and females have shortened oesophageal glands located close to the nerve ring. In the second group of isolates, oesophageal glands are well developed and elongated, which is characteristic of the genus Spilotylenchus. In the first group, the stylet possesses a heavily sclerotized distal spear with a length of approximately half the total stylet length and has a stem with a weaker sclerotization and widening to the base. This stylet structure is characteristic of the genus Rubzovinema, and stylet length (18.5 (14–22) μm) is in accordance with morphometrics given in the description of this genus [26]. In the genus Spilotylenchus, the stylet varies in shape but always possesses a shortened conical distal spear. In the second group of isolates, the stylet structure was similar to that ofSpilotylenchus. Also, the vulval lips of the first group are more protruded than in Spilotylenchus. Other features, including the morphometrics, vary widely in both genera, which hampers taxonomic identification. Nevertheless, based on distinctive traits, we identified the first and second group of isolates as Rubzovinema sp. and Spilotylenchus sp., respectively.

In the genus Rubzovinema, the single species described to date is Rubzovinema ceratophylla [26]. This species is known to parasitize exclusively the flea Citellophilus tesquorum that feeds on sousliks. The specimens of Rubzovinema studied in this work were isolated from five flea species, C. tesquorum, Neopsylla setosa,Frontopsylla semura, Amphipsylla rossica, and Ctenophthalmus secundus, of which the latter two were sampled on mouse-like rodents. Also, the parasitic females of Rubzovinema sp. differed from R. ceratophylla by morphology; they have a shorter tail and more protruded vulval lips. A morphometric comparison ofRubzovinema sp. and R. ceratophylla is given in Table 4.

The parasitic females of Spilotylenchus sp. were isolated from the flea Mesopsylla hebes associated with jerboas. The females were not identified to the species level because of a small number of available specimens and the lack of a free-living stage. A morphometric comparison of Spilotylenchus sp. and the morphologically closest species Spilotylenchus maisonabei [23] is given in Table 5.

In the genus Psyllotylenchus, descriptions of most species are fragmentary and incomplete, which precluded the species identification of the Psyllotylenchus isolates from the fleas N. setosa and F. semura feeding on sousliks. A morphometric comparison of Psyllotylenchus sp. and the type species of this genus, Psyllotylenchus viviparous [25], is given in Table 6.

The 18S and 28S rDNA sequences of Rubzovinema sp. specimens from A. rossica and C. secundus were 100% identical, which indicates that the isolates belong to the same species. The sequences of Rubzovinema sp. ex C. tesquorum, Rubzovinema sp. ex N. setosa, and Rubzovinema sp. ex F. semura diverged from one another and from the gene sequences of Rubzovinema sp. ex A. rossica and Rubzovinema sp. ex C. secundus by 0.4–0.7%, which corresponds to the levels of intraspecific variation [14, 114–119]. The 18S and 28S rDNA sequences ofPsyllotylenchus sp. ex N. setosa and Psyllotylenchus sp. ex F. semura were 100% identical, indicating that they belong to the same species. The 18S and 28S rDNA sequences of Rubzovinema sp. and Psyllotylenchus sp. diverge by 1.2% and 1.9%, respectively. Those of Spilotylenchus sp. ex M. hebes were found to be more divergent. The degree of divergence of the 18S rDNA sequence of Spilotylenchus sp. ex M. hebes from those of either Rubzovinema sp. or Psyllotylenchus sp. was 2.4%; the D3 expansion segment of 28S rDNA diverged by 13.1% and 12.0%, respectively. The observed divergence rate of rDNA sequences agrees well with published evidence on entomoparasitic nematodes [14, 114–118]. Thus, intraspecific divergence of 18S rDNA inDeladenus siricidicola is 1% [120], of D2 and D3 expansion segments in the phytoparasite Bursaphelenchus xylophilus is from 0% to 0.6%, and the interspecific variation between the phytoparasites B. xylophilus andBursaphelenchus mucronatus is from 1.7% to 3.7%. The spacers ITS1 and ITS2 are generally more diverged; the intra- and interspecific variation for these species is from 0 to 3.1% and 11.2 to 13.4%, respectively [121–123].

Molecular vouchering is proved to efficiently complement morphological species identification in nematodes [73, 122, 124–128]. Combining the rDNA and morphological data confirms the species identity within each of the three studied groups of isolates.

Phylogenetic Analysis

In phylogenetic analyses of rDNA we used a dataset with extensive species and gene sampling (SSU-ITS1-5.8S-LSU) compared to earlier published tylenchid phylogenies, most of which were based on SSU rDNA or D2-D3 expansion segments [17, 19–21, 39, 129]. The SSU-ITS1-5.8S-LSU rDNA tree topology (Figure 5) is highly similar to other published phylogenies of tylenchids. In this tree, tylenchomorphs are represented by the sister groups Aphelenchidae and Tylenchida. Most of the tylenchid clades occur in published trees but often contradict classifications based on morphology, as it was also noted by other authors [17, 19–21, 39, 129]. The three robust major branches in the SSU-ITS1-5.8S-LSU rDNA tree (Bayesian posterior probabilities of 0.99–1.0) are (1) the clade includes representatives of the suborders Hoplolaimina, Criconematina, and Tylenchina (excluding Anguinoidea); (2) the majority of classic Anguinata; (3) the suborder Hexatylina. The studied parasites of fleas form a monophyletic group (bootstrap support of 100%) within the Hexatylina.

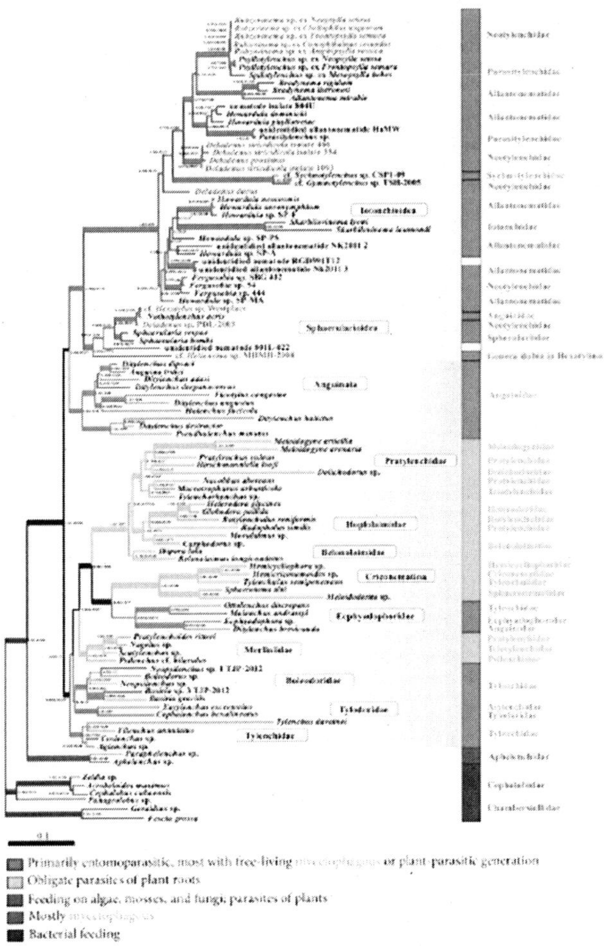

Figure 5: Phylogenetic tree of Tylenchida, inferred from SSU-ITS1-5.8S-LSU rDNA sequences. Topology was inferred using the PhyloBayes software (max-

diff = 0.36). Node support values are shown as follows: the first two values are Bayesian posterior probability assessed using the PhyloBayes and MrBayes software, respectively, and the third is bootstrap support assessed by the ML method. Thick lines lead to the nodes, in which at least one support value of posterior probability is 0.95 and higher. Names of clades (framed) are mainly given by type genera included in them (with the exception of Iotonchioidea). Formal taxonomic position (family by [8]) is shown on the right to the color bar. Colors indicate the ecologies (see the legend). Names of the species of Hexatylina that have a mycetophagous stage in their life cycle are shown in blue. The three robust major branches of Tylenchida are marked by gradient.

The nonredundant rDNA data on the Hexatylina in GenBank mostly represents the D2-D3 expansion segments of LSU rDNA. To maximize species sampling of the Hexatylina, we chose the D3 expansion segment as the molecular marker. The phylogenetic tree with the Anguinoidea as an outgroup is shown in Figure 6. In this tree, the suborder Hexatylina consists of two well-supported clades, in accordance with previously published D2-D3 rDNA phylogenies [19, 20, 39]. The clade of the studied flea parasites is placed within the largest branch of the Hexatylina, similarly to the result of the concatenated rDNA analysis.

The three alternative relationships between the three major branches of Tylenchida (Figure 5) are not discriminated by the AU and Kishino and Hasegawa tests, and only the basal position of the Hexatylina is rejected by the expected-likelihood weights test (Table 7). All three tests do not discriminate between the alternative placement of the flea parasites as closest to the Allantonema, Parasitylenchus, or Deladenus branches; however, its positioning outside this grouping is not rejected only by a less conservative Shimodaira-Hasegawa test [50].

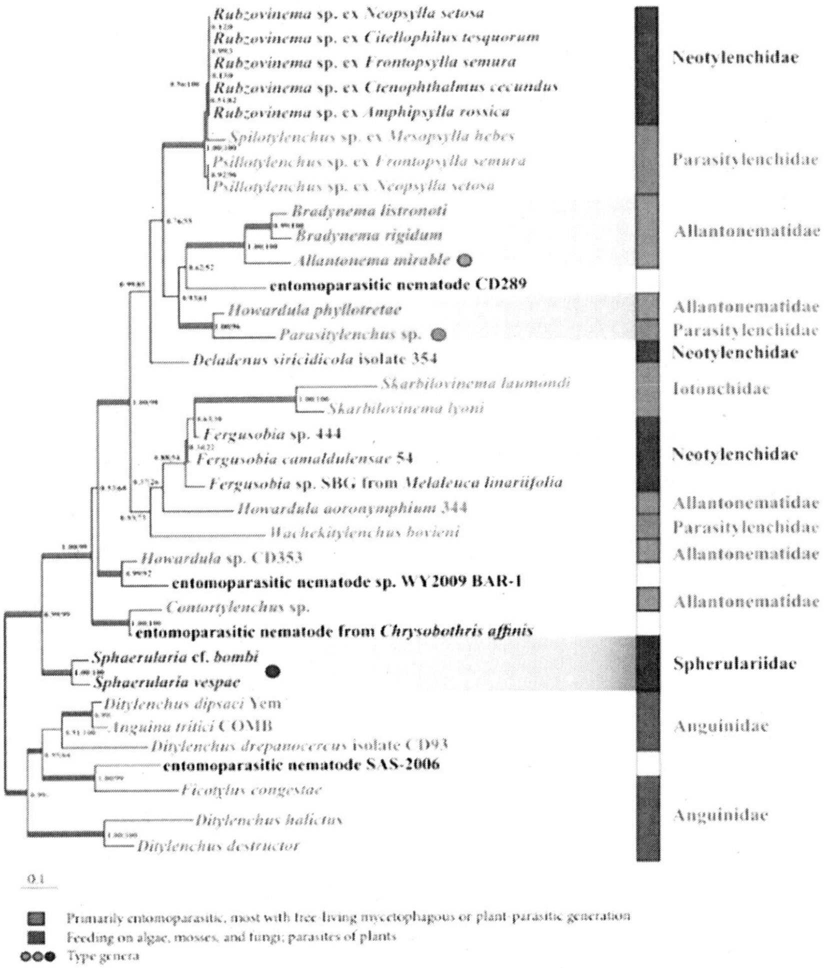

Figure 6: Phylogenetic tree of Hexatylina, inferred from D3 expansion segment of LSU rDNA. Topology was inferred using the PhyloBayes software. Node support values are shown as follows: Bayesian posterior probability/ bootstrap support assessed by the ML method. Thick lines indicate the nodes supported at the level of 0.95 and higher. Color of lines indicates the ecologies (see the legend). Names of species were shown in different colors indicating their taxonomic position. Three families that include their type genera (shown as circles) are marked by gradient.

Table 7: Results of tree topology tests for alternative hypotheses on () the initial divergence of Tylenchida

Topology	Rank	obs	au	np	bp	pp	kh	sh	c-ELW
1									
(((H,An),T),o)	1	−1.8	0.787	0.415	0.402	0.804	0.663	0.969	0.4197
((An,(H,T)),o)	2	4.1	0.326	0.198	0.205	0.013	0.254	0.623	0.1848
((H,(An,T)),o)	3	6.9	0.061	0.013	0.014	0.001	0.101	0.492	0.0186
2									
((((*,Al),P),Ds),o)	1	−1.8	0.787	0.415	0.402	0.804	0.663	0.969	0.4197
((((*,P),Al),Ds),o)	2	1.8	0.495	0.242	0.247	0.130	0.337	0.813	0.2249
(((*,(Al,P)),Ds),o)	3	2.7	0.371	0.110	0.105	0.052	0.243	0.824	0.1209
((*,((Al,P),Ds)),o)	6	15.7	0.063	0.024	0.025		0.053	0.153	0.0272
(((*,Ds),(Al,P)),o)	7	18.3	0.013	0.002	0.002		0.020	0.096	0.0028

Al: Allantonematidae, An: Anguinata, Ds: Deladenus siricidicola—D. proximus group, H: Hexatylina, P: Parasitylenchidae, T: Tylenchina, o: outgroup.

DISCUSSION

Ribosomal DNA Phylogeny of the Tylenchida and Relationships within the Suborder Hexatylina

Phylogenetic analyses of SSU [16, 17, 19, 39] and D2-D3 [20, 39] rDNA data using various methods and species sampling generally agree on the monophyly of most tylenchid clades and contradict classic morphology based classifications. In the SSU-ITS1-5.8S-LSU tree (Figure 5), the monophyletic Tylenchida consists of three major robust clades. The first clade diverges into six groups: (1) the "Tylenchidae (part 2)" (by [17]), (2) the Tylodoridae (represented by the two genera, Cephalenchus and Eutylenchus [83]), (3) Boleodorinae + "Tylenchidae (part 1)" (by [Bert]), (4) the Merliniidae [130], (5) Criconematina + Sphaeronematidae + selected Tylenchina, and (6) Belonolaimidae + "Hoplolaimina." The Merliniidae group corresponds to Clade C in [19] and includes partially the polyphyletic "Telotylenchinae" [131], "Pratylenchidae", and "Hoplolaimina" (Psilenchus cf. hilarulus). Group (5) corresponds to Clade 12A in [129], where Sphaeronematidae (Sphaeronema and Meloidoderita) were earlier shown to be closely related to Criconematina [20, 89], and selected Ecphyadophoridae + Ottolenchus + Malenchus were found to represent a monophyletic clade within the paraphyletic Tylenchina likely to be related to the Criconematina [18, 82]. Group (6) corresponds to Clade VII in [20], Clade 12B in [129], and Clade A + Clade B in [19]. Belonolaimidae (the genera Belonolaimus andIbipora) tend to occupy the basal position. Clade A in [19] contains a "long branch" of the burrowing nematode Radopholus similes ("Pratylenchidae") in sister position to the Hoplolaimidae [17, 19]. This nematode occupies a similar position relative to the Hoplolaimidae in the SSU-ITS1-5.8S-LSU tree, and we consider this unlikely to be an LBA artefact. Similarly to [95], Carphodorus and Morulaimus that belong to the classic Belonolaimidae comprise the basal branch of Clade A sensu [19]. The clade corresponding to Clade B in [19] contains Meloidogynidae, Dolichodoridae, paraphyletic Pratylenchidae, and a part of Telotylenchidae.

The second major clade of the Tylenchida includes representatives of the classic infraorder Anguinata, with a well-supported monophyletic origin, except for a few species. They belong outside the second clade and may initially have been wrongly identified.

The third major clade includes representatives of the classic suborder Hexatylina and consists of two groups. The smaller one unites the three species of Sphaerularia, Helionema sp., cf. Hexatylus sp., Deladenus sp. PDL-2005, and Nothotylenchus acris (Anguinata: Nothotylenchidae). It is further referred to as the Sphaerularioidea according to the type genus. The larger group contains the clade of studied flea parasites and members of the superfamilies Iotonchioidea (Skarbilovinema spp., Parasitylenchus spp., and Wachekitylenchus bovieni) and Sphaerularioidea (Allantonema mirable, Bradynema spp., Howardula spp., and Contortylenchus sp. (fam. Allantonematidae); Deladenus durus, Deladenus proximus, Deladenus siricidicola, Fergusobia spp., andGymnotylenchus sp. (fam. Neotylenchidae)). One species of the Anguinata, Sychnotylenchus sp., also joins the larger group. Our study renders the genera Howardula and Deladenus paraphyletic, as was earlier shown in [19,39, 71, 119].

The genus Howardula is paraphyletic in published rDNA and mitochondrial COI phylogenies [71]. Such characters of Howardula as the degeneration of oesophagus, tail shape, and the absence of stylet in males seem to have evolved independently by convergence. The paraphyletic genus Deladenus is more closely related to either ancestral forms of the Hexatylina or forms typical to the Anguinata. The infraorder Anguinata includes soil-dwelling nematodes, mostly mycetophagous or parasitizing various parts of plants. However, an unidentified entomoparasitic nematode was also grouped within the Anguinoidea [39]. The life cycle ofDeladenus spp. is an irregular alternation of free-living and entomoparasitic forms. The nematode D. siricidicola is able of producing an unlimited number of free-living generations in the absence of the host larvae of siricid pine-killing wood wasps [132]. Like in Anguinata, the free-living forms of Deladenus spp. are fungal feeding. Such characters of Deladenus asthe mycetophagy, enlargement of subventral glands in entomoparasitic females versus their reduction in free-living forms, the hypertrophy of dorsal glands, and stylet reduction in free-living forms seem to be symplesiomorphic. Resemblance with the Anguinata is also typical of other mycetophagous free-living forms: Hexatylus (Neotylenchidae),

Rubzovinema (Neotylenchidae), Prothallonema(Sphaerularioidae) Helionema (Hexatylina dubia), and Paurodontidae. For the latter, the entomoparasitic stage is expected but has never been observed. The relationship between the Hexatylina and Anguinata was earlier hypothesized based on morphology [7, 8, 130, 133, 134]. On rDNA phylogenies of tylenchids, the monophyly of the Hexatylina + Anguinata is either supported [19] or not rejected [20]. In the SSU-ITS1-5.8S-LSUrDNA tree obtained in this study, the monophyly of the Hexatylina + Anguinata has the Bayesian posterior probability of 0.91, but the maximum-likelihood bootstrap support is low; the AU and Kishino and Hasegawa tests did not discriminate between alternative hypotheses.

According to our SSU-ITS1-5.8S-LSU rDNA phylogeny (Figure 5), the major robust branches of the Tylenchida are incongruent with morphology-based classifications suggesting three rather than four suborders (the rank is adopted from morphological systems of tylenchids). Among them, the Hexatylina and Anguinata (both are monophyletic) are likely to be sister groups. The third emerged suborder includes representatives of three classic suborders: Tylenchina, Hoplolaimina, and Criconematina, among which only the latter does not contradict morphology-based classifications.

Considering ecological traits coded in Figure 5, the mycetophagy and/or facultative ectophytoparasitism are likely to be ancestral in the Tylenchida. Sedentary phytoparasites (root-knot species of Meloidogyne, the false root-knot genus Nacobbus, and cyst-forming Heterodera and Globodera) and other obligate endoparasites of plants evolved several times from free-living or facultative sedentary forms, as it was previously hypothesized in accordance with the concept of evolutionary trend to endoparasitism in phytonematodes [135]. Similarly, obligate endoparasites of insects from the Hexatylina are likely to have evolved from mycetophagous forms, with some species retaining the ancestral mycetophageous stage in the life cycle (e.g., species of the paraphyletic genus Deladenus and flea nematodes of the genus Rubzovinema). An interesting specific case in the Hexatylina is the genus Fergusobia that includes plant parasites associated with insects [68, 70], which may have transited to plant parasitism via entomoparasitism [39].

Ribosomal DNA Phylogeny of the Flea Nematodes and Their Classification

The nematodes of fleas do not group with the families known as their relatives in morphology-based systems, as these families do not form monophyletic groups in the tree. However, they do group with both type genera of the families Parasitylenchidae and Allantonematidae (Parasitylenchus and Allantonema, resp.). This grouping is preceded by a successive divergence of Deladenus durus and Deladenus siricidicola (Figure 5). As mentioned above, the pronounced free-living form in Deladenus seems to be ancestral to this group.

Only 31 tylenchid species that parasitize in fleas have been described to date. They differ by morphology, life cycle, and the host specificity, and belong to the five genera: Spilotylenchus (8 species), Psyllotylenchus (20 species), Incurvinema (1 species) Kurochkinitylenchus (1 species), and Rubzovinema (1 species). According to the classification of Siddiqi [8], the genera Spilotylenchus and Psyllotylenchus belong to the family Parasitylenchidae, whereas the genus Rubzovinema is a member of the Neotylenchidae. The two families represent two superfamilies, lotonchioidea and Sphaerularioidea, respectively. All rDNA phylogenies published to date suggest that these superfamilies are paraphyletic [19, 20, 39], which is also inferred in our study with an extensive gene and taxon sampling.

A high degree of rDNA similarity in the three studied species suggests a closer relationship of these species than that assumed by the accepted system of classification. Earlier, Slobodyanyuk proposed to unite all known flea parasites into one family, the Spilotylenchidae. Its four subfamilies, Spilotylenchinae, Rubzovinematinae, Psyllotylenchinae, and Kurochkinitylenchinae, are discriminated based on the life cycle features [28]. In Spilotylenchinae and Rubzovinematinae, the entomoparasitic stage is represented by parasitic females of one heterosexual generation. In Psyllotylenchinae, in addition to the heterosexual generation, a parthenogenetic generation occurs in the flea haemocoel. In Kurochkinitylenchinae, two heterosexual generations exist in the haemocoel: the first generation produces parasitic females and the second generation produces both females and males [28]. Siddiqi also considered the unification of all flea tylenchids into one family but observed the need for further evidence in support [8].

Our results strongly suggest the inclusion of the three genera, Rubzovinema, Psyllotylenchus, and Spilotylenchus, in one family, the Spilotylenchidae [28]. The ribosomal DNA genetic distance within the family Spilotylenchidae is much smaller than that of certain tylenchid genera, for example, Meloidogyne (Figure 4) orPratylenchus [19, 84].

Host Specificity of Flea Nematodes

The majority of tylenchid nematodes are monoxenous or oligoxenous; in particular, flea parasites were thought to be strictly host specific. Earlier papers suggested the lack of strict host specificity in Psyllotylenchus pawlowskyi and Psyllotylenchus viviparous [13, 25]. However, later these species were found to be heterogeneous and sustained revision [9, 27–29]. Spilotylenchus pawlowskyi and Spilotylenchus caspius were referred to as single-host parasites of the flea Coptopsylla lamellifer [27, 136]. Kurochkinitylenchus laevicepsi andSpilotylenchus ivashkini also share the same flea host, Nosopsyllus laeviceps [28, 29]. Before our study, the genusRubzovinema was known to contain a single species, Rubzovinema ceratophylla, which parasitizes exclusively the flea Citellophilus tesquorum.

We found that at least two out of the three studied species are not single-host parasites. Psyllotylenchus sp. was shown to parasitize two flea species feeding on sousliks, Frontopsylla semura and Neopsylla setosa.Rubzovinema sp. was found on five flea species feeding on different rodent hosts: C. tesquorum, F. semura, N. setosa (all sampled from sousliks), Ctenophtalamus secundus, and Amphipsylla rossica (all sampled from voles).A. rossica, F. semura, and C. tesquorum belong to different families of the superfamily Ceratophylloidea (Leptopsyllidae and Ceratophyllidae), whereas C. secundus and N. setosa belong to the superfamily Hystrichopsylloidea. Unlike the host-specific R. ceratophylla, the studied Rubzovinema sp. parasitizes taxonomically distant fleas feeding on different rodents. Thus, the common opinion that flea nematodes are strictly host specific should be revisited.

As the two species of Rubzovinema demonstrate, even closely related parasites may exhibit different host range size. Among other known examples are the entomoparasitic nematodes of the genus Howardula parasitizing various beetles and flies [71, 137, 138], many phytonematodes [8], sibling species of parasitoid flies [128], and herbivorous insects [139]. The host range of parasites is an indicator of their evolutionary strategy in the ecosystem. Multihost parasites can be considered ecological generalists, in contrast to specialists that coevolve with a particular host. Generalists and specialists play different roles in the ecosystem [140], where they keep in balance, taking advantages and disadvantages of the two strategies. The advantages of generalization are yet to be explained by evolutionary biologists, whereas advantages of specialization are obvious, and it is generally accepted that evolution favors specialism [141, 142]. In the flea parasites, this trend is demonstrated by a greater species diversity of ecological specialists, the genera Spilotylenchus and Psyllotylenchus.

Nevertheless, the generalist Rubzovinema sp. was most abundant in the studied samples, which indicates that extending the host range may be evolutionarily successful. Besides the need to combat the immune response of several hosts, which is a requirement to widen the hosts range [143], the free-living stage of Rubzovinema sp. is to adapt to diverse microbioclimatic conditions of complex environments of rodent habitats. Multihost parasites pay a cost of adapting to alternative conditions [141, 144] compensated by stable survival of the species. Considering the spatial and temporal dynamics of flea populations feeding on a particular rodent host (one or two flea species usually dominate over a sampling season), multihost nematode parasites gain an advantage of their relative independence of population waves of either flea hosts or their rodent hosts. A higher infestation rate observed for Rubzovinema sp., compared to the two other studied species, may be an indicator of a greater ecological plasticity of this multihost parasite.

Entomoparasitic Nematodes in Natural Foci of Plague

In natural foci of plague, the epizootic dynamics are influenced by numerous climatic and biotic factors. The spatial and temporal population dynamics of the plague agent, Y. pestis, affect the population dynamics of the flea vectors and their mammalian hosts. Members of the transmission route of the plague agent also closely interact with other living organisms. For example, parasites of fleas that in turn feed on rodents are hyperparasites that play the role of high-level control agents on the ecosystem level, the role that entomoparasitic nematodes share with the bacterial plague agent. High-level control agents render the epidemiological state of a natural focus of disease less predictable. On the one hand, a lower density of the flea vector population reduces the plaque transmission rate; on the other, its growth causes an exponential decay of the host rodent population [145] below its epidemiological threshold, above which there is a threat of spillover of plague infection into human population [145]. Hypothetically, nematode-induced decrease of flea population is able to increase the number of rodents above the threshold and thus trigger an epidemic. The dual effect of high-level control agents is well exemplified by cases, when during plague episodes the extermination of rodents by humans causes the return of infection through stimulating the migration of fleas, the plaque vectors [5].

The studied entomoparasitic nematodes possess high potential as control agents of the flea vectors of plague owing to their high propagation rate within the flea host (Figure 2) and high infestation level (up to 21% observed in this study and from 50 to 60%, as estimated by other authors [10, 11]). One of the studied nematode species, Rubzovinema sp., is a multihost parasite. Host-specific parasites reach the optimal level of pathogenicity by maintaining the trade-off between pathogenicity and transmissibility. Adding of a new host to a multihost system makes the model more complicated [141]. The multihost parasite Rubzovinema sp. is expected to exhibit different levels of pathogenicity with respect to different flea hosts which, in turn, play different roles in the transmission of plague. Epizootics cause sporadic mortality in local populations of all members involved in the interaction with the plague agent, and their survival is contingent on migrations within a metapopulation. It is the case when the Cope's law [139, 146] governs the extinction of specialists on a shorter time scale rather than a geological period, and evolution may favor the ecological generalists, such asRubzovinema sp.

Some authors surmised the involvement of entomoparasitic nematodes in the transmission of the plague agent [4], as it was observed that biofilms of Yersinia pestis adhere to cuticle receptors of Caenorhabditis elegans[147–149]. In this perspective, nematodes parasitizing fleas in natural foci of plague take on greater importance, as they may provide for the transmission route that does not include a mammal [4]. Further studies will clarify the role of flea nematodes in the transmission of plague infection.

ACKNOWLEDGMENTS

The authors thank G. S. Mirzaeva for help with PCR amplification and rDNA fragment analysis, N. V. Popov and the staff of the Laboratory of Epizootics Monitoring for advice and assistance in collecting and processing rodent samples, and, particularly, A. N. Porshakov for help in identification of flea specimens. They also thank O. V. Slobodyanyuk for helpful discussions of results, S. E. Spiridonov for advice on cultivation of entomoparasitic nematodes, S. A. Subbotin for valuable comments on the earlier version of the paper, and E. Yu. Talanova and L. Yu. Rusin for discussions of its final version, and E. A. Musatkina for assistance with the manuscript preparation They are grateful to the Supercomputer Center of Moscow State University (http://parallel.ru/cluster) and the Bioportal of the University of Oslo (http://www.bioportal.uio.no) for providing computing resources.

REFERENCES

1. A. W. Bacot and C. J. Martin, "Observations on the mechanism of the transmission of plague by fleas,"The Journal of Hygiene, vol. 13, supplement 3, pp. 423–439, 1914.
2. V. V. Kutyrev, G. A. Eroshenko, N. V. Popov, N. A. Vidyaeva, and N. P. Konnov, "Molecular mechanisms of interaction of the plague agent with invertebrates," Molecular Genetics, Microbiology and Virology, vol. 24, no. 4, pp. 7–12, 2009.
3. E. N. Pavlovsky, Natural Focality of Transmissive Diseases in Connection with Landscape Epidemiology of Zooanthroponoses, Moscow, Russia, 1964.
4. N. V. Popov, E. I. Koshel, G. A. Eroshenko, and V. V. Kutyrev, "Formation of modern concepts on the mechanism of plague enzooty," Problems of Particularly Dangerous Infections, vol. 3, no. 109, pp. 5–8, 2011.
5. M. J. Keeling and C. A. Gilligan, "Bubonic plague: a metapopulation model of a zoonosis," Proceedings of the Royal Society B, vol. 267, no. 1458, pp. 2219–2230, 2000.
6. R. A. Bedding, R. J. Akhurst, and H. K. Kaya, Nematodes and the Biological Control of Insect Pests, CSIRO Press, Melbourne, Australia, 1993.

7. M. R. Siddiqi, Tylenchida Parasites of Plants and Insects, vol. 645 of Commonwealth Agricultural Bureaux, Farnham Royal, Slough, UK, 1986.
8. M. R. Siddiqi, Tylenchida: Parasites of Plants and Insects, CABI, Wallingford, UK, 2nd edition, 2000.
9. I. A. Rubtsov, Parasites and Enemies of Fleas, Nauka, Leningrad, Russia, 1981.
10. Y. A. Morozov, "About infestation with fleas great gerbils different ages," in Proceedings of the Conference Anti-Plague Facilities in Central Asia and Kazakhstan, pp. 337–338, Alma-Ata, 1974.
11. Y. A. Morozov, "Effect of infestation of nematode on reproduction of fleas gerbils in Muyunkum," inProceedings of the Conference Anti-Plague Facilities in Central Asia and Kazakhstan, pp. 338–340, Alma-Ata, 1974.
12. J. Deunff, Parasites de Siphonaptères. Étude de la systématique, de la biologie et du pouvoir pathogène des Tylenchides (Nematodea) dans une perspective de lutte biologique [Ph.D. thesis], Etat Sc. Pham, Rennes, France, 1984.
13. Y. V. Kurochhkin, "The nematode Heterotylenchus pawlowskyi sp. n., castrating flea-vectors of plague,"Doklady Akademyi Nauk SSSR, vol. 135, pp. 1281–1284, 1960.
14. M. L. Blaxter, P. De Ley, J. R. Garey et al., "A molecular evolutionary framework for the phylum Nematoda," Nature, vol. 392, no. 6671, pp. 71–75, 1998.
15. P. De Ley and M. L. Blaxter, "Systematic Position and Phylogeny," in The Biology of Nematodes, D. L. Lee, Ed., Taylor & Francis, London, UK, 2002.
16. B. H. M. Meldal, N. J. Debenham, P. De Ley et al., "An improved molecular phylogeny of the Nematoda with special emphasis on marine taxa," Molecular Phylogenetics and Evolution, vol. 42, no. 3, pp. 622–636, 2007.
17. W. Bert, F. Leliaert, A. R. Vierstraete, J. R. Vanfleteren, and G. Borgonie, "Molecular phylogeny of the Tylenchina and evolution of the female gonoduct (Nematoda: Rhabditida)," Molecular Phylogenetics and Evolution, vol. 48, no. 2, pp. 728–744, 2008.
18. M. Holterman, A. Van Der Wurff, S. Van Den Elsen et al., "Phylum-wide analysis of SSU rDNA reveals deep phylogenetic

relationships among nematodes and accelerated evolution toward crown clades,"Molecular Biology and Evolution, vol. 23, no. 9, pp. 1792–1800, 2006. View at Publisher ·

19. M. Holterman, G. Karssen, S. Van Den Elsen, H. Van Megen, J. Bakker, and J. Helder, "Small subunit rDNA-based phylogeny of the tylenchida sheds light on relationships among some high-impact plant-parasitic nematodes and the evolution of plant feeding," Phytopathology, vol. 99, no. 3, pp. 227–235, 2009.
20. S. A. Subbotin, D. Sturhan, V. N. Chizhov, N. Vovlas, and J. G. Baldwin, "Phylogenetic analysis of Tylenchida Thorne, 1949 as inferred from D2 and D3 expansion fragments of the 28S rRNA gene sequences," Nematology, vol. 8, no. 3, pp. 455–474, 2006r
21. S. A. Subbotin, D. Sturhan, B. J. Adams, et al., "Molecular phylogeny of the order Tylenchida: analysis of nuclear ribosomal RNA genes," Journal of Nematology, vol. 38, no. 2, pp. 296–296, 2006.
22. H. Launay, J. Deunff, and O. Bain, "Spilotylenchus arthuri, gen. n., sp. n. (Nematodea, Tylenchida: Allantonematidae), parasite of Spilopsyllus cuniculi (Dale, 1878) (Siphonaptera: Pulicidae)," Annales de Parasitologie Humaine et Comparee, vol. 58, no. 2, pp. 141–150, 1983.
23. H. Launay and J. Deunff, "Spilotylenchus maisonabei n. sp. (Nematoda: Allantonematidae) parasite of the European rabbit flea Spilopsyllus cuniculi (Dale, 1878) (Siphonaptera: Pulicidae)," Annales de Parasitologie Humaine et Comparee, vol. 13, pp. 293–296, 1990.
24. C. Laumond and J. C. Beaucournu, "Neoparasitylenchus megabothridis n. sp. (Tylenchida: Allantonematidae) parasite of Megabothris tubidus (Siphonaptera: Ceratophyllidae); observations on fleas tylenchides in the S.W. of Europa (author›s transl)," Annales de Parasitologie Humaine et Comparee, vol. 53, no. 3, pp. 291–302, 1978.
25. G. O. Poinar Jr. and B. C. Nelson, "Psyllotylenchus viviparus, n. gen., n. sp. (Nematodea: Tylenchida: Allantonematidae) parasitizing fleas (Siphonaptera) in California," Journal of Medical Entomology, vol. 10, no. 4, pp. 349–354, 1973.
26. V. O. Slobodyanyuk, "Validation of genus Rubzovinema gen.n. (Sphaerularioidea) and redescription of species Rubzovinema

ceratophylla comb.n. parasite of fleas Citellophillus tesquorum," Zoological Journal, vol. 70, no. 9, pp. 33–43, 1991.

27. O. V. Slobodyanyuk, "Revision of the species Psyllotylenchus pawlowskyi (Kurochkin, 1960) Poinar & Nelson, 1973. I. Redescription of Spilotylenchus pawlowskyi (sensu stricto) comb. n. (Tylenchida: Allantonematidae)," Russian Journal of Nematology, vol. 5, no. 2, pp. 103–112, 1997
28. O. V. Slobodyanyuk, "Revision of the species Psyllotylenchus pawlowskyi (Kurochkin, 1960) Poinar & Nelson, 1973. II. Description of Kurochkinitylenchus laevicepsi gen. n., sp. n. and Spilotylenchidae fam. n," Russian Journal of Nematology, vol. 7, no. 1, pp. 1–18, 1999.
29. O. V. Slobodyanyuk, "Revision of the species Psyllotylenchus pawlowskyi (Kurochkin, 1960) Poinar & Nelson, 1973. III. Description of Spilotylenchus ivashkini sp. n," Russian Journal of Nematology, vol. 8, no. 1, pp. 45–55, 2000.
30. J. Deunff and H. Launay, "Psyllotylenchus chabaudi, n. sp. (Nematodea, Tylenchida: Allantonematidae), a parasite of Nosopsyllus fasciatus (Bosc) (Siphonaptera: Ceratophyllidae)," Annales de Parasitologie Humaine et Comparee, vol. 59, no. 3, pp. 263–270, 1984.
31. J. Deunff, H. Launay, and J. C. Beaucournu, "Incurvinema helicoides n. gen., n. sp. Nematodea, Tylenchida: allantonematidae parasite de Rhadinopsylla pentacantha (Rothschild, 1897) (Siphonaptera : Hystrichopsyllidae)," Annales de Parasitologie Humaine Et Comparée, vol. 60, pp. 739–746, 1985.
32. A. T. De Grisse, "Redescriptions ou modifications de quelques techniques utilisées dans l›étude des nématodes phytoparasites," Mededelingen Rijksfakulteit Landbouwwetenschappen Gent, vol. 34, pp. 351–369, 1969.
33. X. Huang, "A contig assembly program based on sensitive detection of fragment overlaps," Genomics, vol. 14, no. 1, pp. 18–25, 1992
34. T. A. Hall, "BioEdit: a user-friendly biological sequence alignment editor and analysis program for Windows 95/98/NT," Nucleic Acids Symposium Series, vol. 41, pp. 95–98, 1999.

35. L. Medlin, H. J. Elwood, S. Stickel, and M. L. Sogin, "The characterization of enzymatically amplified eukaryotic 16S-like rRNA-coding regions," Gene, vol. 71, no. 2, pp. 491–499, 1988.
36. V. N. Chizhov, O. A. Chumakova, S. A. Subbotin, and J. G. Baldwin, "Morphological and molecular characterization of foliar nematodes of the genus Aphelenchoides: A. fragariae and A. ritzemabosi(Nematoda: Aphelenchoididae) from the Main Botanical Garden of the Russian Academy of Sciences, Moscow," Russian Journal of Nematology, vol. 14, no. 2, pp. 179–184, 2006.
37. G. Van der Auwera, S. Chapelle, and R. De Wächter, "Structure of the large ribosomal subunit RNA ofPhytophthora megasperma, and phylogeny of the oomycetes," FEBS Letters, vol. 338, no. 2, pp. 133–136, 1994.
38. S. A. Subbotin, N. Vovlas, R. Crozzoli, et al., "Phylogeny of Criconematina Siddiqi, 1980 (Nematoda: Tylenchida) based on morphology and D2-D3 expansion segments of the 28S rDNA gene sequences with application of a secondary structure model," Nematology, vol. 7, pp. 927–944, 2005.
39. V. N. Chizhov, N. N. Butorina, and S. A. Subbotin, "Entomoparasitic nematodes of the genus Skarbilovinema: S. laumondi and S. lyoni (Nematoda: Tylenchida), parasites of the flies of the family Syrphidae (Diptera), with phylogeny of the suborder Hexatylina," Russian Journal of Nematology, vol. 20, no. 2, pp. 141–155, 2012.
40. S. F. Altschul, T. L. Madden, A. A. Schäffer et al., "Gapped BLAST and PSI-BLAST: a new generation of protein database search programs," Nucleic Acids Research, vol. 25, no. 17, pp. 3389–3402, 1997
41. R. C. Edgar, "MUSCLE: multiple sequence alignment with high accuracy and high throughput," Nucleic Acids Research, vol. 32, no. 5, pp. 1792–1797, 2004.
42. M. Nei and S. Kumar, Molecular Evolution and Phylogenetics, Oxford University Press, Oxford, UK, 2000.
43. N. Lartillot, T. Lepage, and S. Blanquart, "PhyloBayes 3: a Bayesian software package for phylogenetic reconstruction and molecular dating," Bioinformatics, vol. 25, no. 17, pp. 2286–2288, 2009.

44. N. Lartillot and H. Philippe, "A Bayesian mixture model for across-site heterogeneities in the amino-acid replacement process," Molecular Biology and Evolution, vol. 21, no. 6, pp. 1095–1109, 2004.
45. J. P. Huelsenbeck and F. Ronquist, "MRBAYES: bayesian inference of phylogenetic trees,"Bioinformatics, vol. 17, no. 8, pp. 754–755, 2001.
46. C. Lanave, G. Preparata, C. Saccone, and G. Serio, "A new method for calculating evolutionary substitution rates," Journal of Molecular Evolution, vol. 20, no. 1, pp. 86–93, 1984.
47. A. Stamatakis, "Maximum likelihood-based phylogenetic analyses with thousands of taxa and mixed models," Bioinformatics, vol. 22, pp. 2688–2690, 2006.
48. H. Shimodaira, "An approximately unbiased test of phylogenetic tree selection," Systematic Biology, vol. 51, no. 3, pp. 492–508, 2002
49. H. Kishino and M. Hasegawa, "Evaluation of the maximum likelihood estimate of the evolutionary tree topologies from DNA sequence data, and the branching order in Hominoide," Journal of Molecular Evolution, vol. 29, no. 2, pp. 170–179, 1989.
50. H. Shimodaira and M. Hasegawa, "CONSEL: for assessing the confidence of phylogenetic tree selection,"Bioinformatics, vol. 17, no. 12, pp. 1246–1247, 2002.
51. K. Strimmer and A. Rambaut, "Inferring confidence sets of possibly misspecified gene trees," Proceedings of the Royal Society B, vol. 269, no. 1487, pp. 137–142, 2002
52. H. A. Schmidt, K. Strimmer, M. Vingron, and A. Von Haeseler, "TREE-PUZZLE: maximum likelihood phylogenetic analysis using quartets and parallel computing," Bioinformatics, vol. 18, no. 3, pp. 502–504, 2002.
53. R. D. M. Page, "TreeView: an application to display phylogenetic trees on personal computers,"Computer Applications in the Biosciences, vol. 12, no. 4, pp. 357–358, 1996.
54. O. Holovachov, S. Bostrom, and I. T. De Ley, et al., "Morphology, molecular characterisation and systematic position of the genus Cynura Cobb, 1920 (Nematoda: Plectida)," Nematology, vol. 15, pp. 611–627, 2013.

55. S. A. Nadler, P. De Ley, M. and Mundo-Ocampo et al., "Phylogeny of Cephalobina (Nematoda): molecular evidence for recurrent evolution of probolae and incongruence with traditional classifications,"Molecular Phylogenetics and Evolution, vol. 40, no. 3, pp. 696–711, 2006.
56. O. Holovachov, S. Bostrom, S. A. Nadler, and P. De Ley, "Systematics and phylogenetic position of the genus Tricirronema Siddiqi, 1993 (Cephalobomorpha)," Journal of Nematode Morphology and Systematics, vol. 12, no. 2, 2009.
57. K. Kiontke, A. Barrière, I. Kolotuev et al., "Trends, stasis, and drift in the evolution of nematode vulva development," Current Biology, vol. 17, no. 22, pp. 1925–1937, 2007
58. R. Campos-Herrera, F. E. El-Borai, and L. W. Duncan, "Wide interguild relationships among entomopathogenic and free-living nematodes in soil as measured by real time qPCR," Journal of Invertebrate Pathology, vol. 111, no. 2, pp. 126–135, 2012.
59. M.-A. Félix, P. De Ley, and R. J. Sommer et al., "Evolution of vulva development in the Cephalobina (Nematoda)," Developmental Biology, vol. 221, no. 1, pp. 68–86, 2000
60. S. Boström, O. Holovachov, and S. A. Nadler, "Description of Scottnema lindsayae Timm, 1971 (Rhabditida: Cephalobidae) from Taylor Valley, Antarctica and its phylogenetic relationship," Polar Biology, vol. 34, no. 1, pp. 1–12, 2011.
61. J. W. Waceke, D. J. Bumbarger, M. Mundo-Ocampo, S. A. Subbotin, and J. G. Baldwin, "Zeldia spannata n. sp. (Nematoda: Cephalobidae) from the Mojave Desert, California," Journal of Nematode Morphology and Systematics, vol. 8, no. 1, pp. 57–67, 2005.
62. S. Kumari, "Aphelenchus avenae (Nematoda: Aphelenchidae) under the rhizosphere of Brassica napus,"Helminthologia, vol. 49, no. 1, pp. 57–59, 2012.
63. H. Iwahori, K. Tsuda, N. Kanzaki, K. Izui, and K. Futai, "PCR-RFLP and sequencing analysis of ribosomal DNA of Bursaphelenchus nematodes related to pine wilt disease," Fundamental and Applied Nematology, vol. 21, no. 6, pp. 655–666, 1998.
64. L. K. Carta, A. M. Skantar, Z. A. Handoo, and M. A. Baynes, "Supplemental description ofParaphelenchus acontioides

(Tylenchida: Aphelenchidae, Paraphelenchinae), with ribosomal DNA trees and a morphometric compendium of female Paraphelenchus," Nematology, vol. 13, no. 8, pp. 887–899, 2011.

65. Y. Zeng, R. M. Giblin-Davis, Y. Weimin, G. Bélair, G. Boivin, and K. W. Thomas, "Bradynema listronotin. sp. (Nematoda: Allantonematidae), a parasite of the carrot weevil Listronotus oregonensis (Coleoptera: Curculionidae) in Quebec, Canada," Nematology, vol. 9, no. 5, pp. 609–623, 2007.
66. K. Rybarczyk-Mydłowska, P. Mooyman, H. van Megen, and Setal, "Small subunit ribosomal DNA-based phylogenetic analysis of foliar nematodes (Aphelenchoides spp.) and their quantitative detection in complex DNA backgrounds," Phytopathology, vol. 102, no. 12, pp. 1153–1160, 2012.
67. Q. Yu, P. de Groot, I. Leal, C. Davis, W. Ye, and B. Foord, "First report and characterization ofDeladenus proximus (Nematoda: Neotylenchidae) associated with Sirex nigricornis (Hymenoptera: Siricidae) in Canada," International Journal of Nematology, vol. 21, no. 2, pp. 139–146, 2012.
68. W. Ye, R. M. Giblin-Davis, and K. A. Davies et al., "Molecular phylogenetics and the evolution of host plant associations in the nematode genus Fergusobia (Tylenchida: Fergusobiinae)," Molecular Phylogenetics and Evolution, vol. 45, no. 1, pp. 123–141, 2007
69. I. Leal, B. Foord, C. Davis, P. de Groot, X. O. Mlonyeni, and B. Slippers, "Distinguishing isolates ofDeladenus siricidicola ,a biological control agent of Sirex noctilio, from North America and the Southern Hemisphere using PCR- RFLP," Canadian Journal of Forest Research, vol. 42, pp. 1173–1177, 2012.
70. K. A. Davies, W. Ye, R. M. Giblin-Davis, G. S. Taylor, S. Scheffer, and W. K. Thomas, "The nematode genus Fergusobia (Nematoda: Neotylenchidae): molecular hylogeny, descriptions of clades and associated galls, host plants and Fergusonina fly larvae," Zootaxa, no. 2633, pp. 1–66, 2010.
71. S. J. Perlman, G. S. Spicer, D. Dewayne Shoemaker, and J. Jaenike, "Associations between mycophagousDrosophila and their Howardula nematode parasites: a worldwide phylogenetic shuffle," Molecular Ecology, vol. 12, no. 1, pp. 237–249, 2003.

72. C. L. Raak-van den Berg, P. S. van Wielink, and P. W. de Jong, et al., "Invasive alien species under attack: natural enemies of Harmonia axyridis in the Netherlands," BioControl. In press.
73. N. Kanzaki, R. M. Giblin-Davis, and R. H. Scheffrahn, et al., "Reverse taxonomy for elucidating diversity of insect-associated nematodes: a case study with termites," PLoS ONE, vol. 7, no. 8, Article ID e43865, 2012.
74. T. O. Powers, D. A. Neher, and P. Mullin et al., "Tropical nematode diversity: vertical stratification of nematode communities in a Costa Rican humid lowland rainforest," Molecular Ecology, vol. 18, no. 5, pp. 985–996, 2009.
75. C. Hazir, R. M. Giblin-Davis, and N. Keskin et al., "Diversity and distribution of nematodes associated with wild bees in Turkey," Nematology, vol. 12, no. 1, pp. 65–80, 2010.
76. M. Holterman, S. van den Elsen, and H. van Megen, et al., "Evolutionary relationships between members of the phylum Nematoda based on small subunit ribosomal DNA sequences," 2004.
77. N. Kanzaki, H. Kosaka, K. Sayama, J.-I. Takahashi, and S. Makino, "Sphaerularia vespae sp. nov. (Nematoda, Tylenchomorpha, Sphaerularioidea), an endoparasite of a common Japanese hornet, Vespa simillima Smith (Insecta, Hymenoptera, Vespidae)," Zoological Science, vol. 24, no. 11, pp. 1134–1142, 2007
78. Q. Yu, "Ditylenchus destructor Thorne, 1945 (Tylenchida: Anguinidae) in Canada," Journal of Nematology, vol. 44, no. 4, pp. 500–500, 2012.
79. B. Steven, L. V. Gallegos-Graves, C. Yeager, J. Belnap, and C. R. Kuske, "Common and distinguishing features of the bacterial and fungal communities in biological soil crusts and shrub root zone soils," Soil Biology and Biochemistry, vol. 69, pp. 302–312, 2014.
80. R. D. L. Oliveira, A. M. Santin, and D. J. Seni, et al., "Ditylenchus gallaeformans sp. n. (Tylenchida: Anguinidae)—a neotropical gall forming nematode with biocontrol potential against weedy Malastomataceae," Nematology, vol. 15, no. 2, pp. 179–196, 2013.
81. K. A. Davies, W. Ye, R. M. Giblin-Davis, and K. W. Thomas, "Ficotylus congestae gen. n., sp. n. (Anguinata), from ficus

congesta (moraceae) sycones in Australia," Nematology, vol. 11, no. 1, pp. 63–75, 2009

82. M. R. Atighi, E. Pourajam, T. J. Pereira, et al., "Redescription of Filenchus annulatus (Siddiqui & Khan, 1983) Siddiqi, 1986 based on specimens from Iran with contributions to the molecular phylogeny of the Tylenchida," Nematology, vol. 15, no. 2, pp. 129–141, 2013.
83. J. E. Palomares-Rius, S. A. Subbotin, G. Liébanas, B. B. Landa, and P. Castillo, "Eutylenchus excretoriusEbsary & Eveleigh, 1981 Nematoda: Tylodorinae) from Spain with approaches to olecular phylogeny of related genera," Nematology, vol. 11, no. 3, pp. 343–354, 2009.
84. S. A. Subbotin, E. J. Ragsdale, T. Mullens, P. A. Roberts, M. Mundo-Ocampo, and J. G. Baldwin, "A phylogenetic framework for root lesion nematodes of the genus Pratylenchus (Nematoda): evidence from 18S and D2-D3 expansion segments of 28S ribosomal RNA genes and morphological characters,"Molecular Phylogenetics and Evolution, vol. 48, no. 2, pp. 491–505, 2008.
85. Z. Majd Taheri, Z. Tanha Maafi, S. A. Subbotin, E. Pourjam, and A. Eskandari, "Molecular and phylogenetic studies on Pratylenchidae from Iran with additional data on Pratylenchus delattrei,Pratylenchoides alkani and two unknown species of Hirschmanniella and Pratylenchus," Nematology, vol. 15, pp. 633–651, 2013.
86. C. Xu, H. Xie, C. Zhao, S. Zhang, and X. Su, "Review of the genus Scutylenchus Jairajpuri, 1971 (Nematoda: Tylenchida), with description of Scutylenchus dongtingensis n. sp. from rhizosphere soil of grass in China," Zootaxa, vol. 3437, pp. 32–42.
87. Y. Yang and S. Zhang, "Identification of Hemicriconemoides gaddi from the Rhizosphere of Longan,"Chinese Journal of Tropical Crops, vol. 5, pp. 935–941, 2013.
88. E. Van den Berg, S. A. Subbotin, and L. R. Tiedt, "Morphological and molecular characterisation ofHemicycliophora lutosa Loof & Heyns, 1969 and H. typica de Man, 1921 from South Africa (Nematoda: Hemicycliophoridae)," Nematology, vol. 12, no. 2, pp. 303–308,
89. N. Vovlas, B. B. Landa, G. Liébanas, Z. A. Handoo, S. A. Subbotin, and P. Castillo, "Characterization of the cystoid nematode

Meloidoderita kirjanovae (Nemata: Sphaeronematidae) from Southern Italy,"Journal of Nematology, vol. 38, no. 3, pp. 376–382, 2006

90. J. E. Palomares-Rius, N. Vovlas, and S. A. Subbotin et al., "Molecular and morphological characterisation ofSphaeronema alni Turkina & Chizhov, 1986 (Nematoda: Sphaeronematidae) from Spain compared with a topotype population from Russia," Nematology, vol. 12, no. 4, pp. 649–659, 2010.
91. G. Liu, J. Chen, S. Xiao, D. Pan, and S. Zhang, "Intraspecific variability of Tylenchulus semipenetranspopulations on citrus and Chinese fir," Scientia Agricultura Sinica, vol. 9, 2011.
92. B. Y. Park, S. N. Park, J. K. Lee, and C. H. Bae, "Morphometric and genetic variability amongTylenchulus semipenetrans populations from citrus growing area in Korea," Plant Pathology Journal, vol. 25, no. 3, pp. 236–240, 2009.
93. U. Gozel, B. J. Adams, K. B. Nguyen, R. N. Inserra, R. M. Giblin-Davis, and L. W. Duncan, "A phylogeny of Belonolaimus populations in Florida inferred from DNA sequences," Nematropica, vol. 36, no. 2, pp. 155–171, 2006.
94. Z. A. Handoo, A. M. Skantar, and P. Mulrooney, "First report of the sting nematode Belonolaimus longicaudatus on soybean in Delaware," Plant Disease, vol. 94, no. 1, p. 133, 2010
95. G. R. Stirling, A. M. Stirling, and R. M. Giblin-Davis, et al., "Distribution of southern sting nematode,Ibipora lolii (Nematoda: Belonolaimidae), on turfgrass in Australia and its taxonomic relationship to other belonolaimids," Nematology, vol. 15, pp. 401–415, 2013.
96. J. Parkinson, M. Mitreva, N. Hall, M. Blaxter, and J. P. McCarter, "400000 nematode ESTs on the Net,"Trends in Parasitology, vol. 19, no. 7, pp. 283–286, 2003.
97. D. D. Sui, N. Atibalentja, G. R. Noel, and L. L. Domier, "Genetic diversity of the rDNA locus among 27 populations and 8 races of Heterodera glycines from China, Japan, and the United States, as revealed by PCR-RFLP," Journal of Nematology. In press.
98. B. Gao, R. Allen, T. Maier, E. L. Davis, T. J. Baum, and R. S. Hussey, "Identification of putative parasitism genes expressed in the esophageal gland cells of the soybean cyst nematode Heterodera

glycines," Molecular Plant-Microbe Interactions, vol. 14, no. 10, pp. 1247–1254, 2001.

99. D. P. Puthoff, M. L. Ehrenfried, B. T. Vinyard, and M. L. Tucker, "GeneChip profiling of transcriptional responses to soybean cyst nematode, Heterodera glycines, colonization of soybean roots," Journal of Experimental Botany, vol. 58, no. 12, pp. 3407–3418, 2007.

100. G. E. Múnera Uribe, W. Bert, A. R. Vierstraete, E. de la Peña, M. Moens, and W. Decraemer, "Burrowing nematodes from Colombia and their relationship with Radopholus similis populations, R. arabocoffeae and R. duriophilus," Nematology, vol. 12, pp. 619–629, 2010

101. J. Jacob, M. Mitreva, B. Vanholme, and G. Gheysen, "Exploring the transcriptome of the burrowing nematode Radopholus similis," Molecular Genetics and Genomics, vol. 280, no. 1, pp. 1–17, 2008.

102. J. Li, D. Peng, and W. Huang, "Phylogentic analysis of Radopholus similis from D2 and D3 fragments of the 28S rRNA gene sequences," Journal of Huazhong Agricultural University, vol. 5, 2008.

103. S. T. Nyaku, V. R. Sripathi, R. V. Kantety, Y. Q. Gu, K. Lawrence, and G. C. Sharma, "Characterization of the two intra-individual sequence variants in the 18S rRNA gene in the plant parasitic nematode,Rotylenchulus reniformis," PLoS ONE, vol. 8, no. 4, Article ID E60891, 2013.

104. Y. Zhan, A. Matafeo, H. Shi, and J. Zheng, "Morphological and molecular characterization and host range of Rotylenchulus reniformis population occurring in Hangzhou, Zhejiang, China," Acta Phytopathologica Sinica, vol. 41, no. 1, pp. 37–43, 2011.

105. Y. Zeng, R. M. Giblin-Davis, and W. Ye, "Two new species of Schistonchus (Nematoda: Aphelenchoididae) associated with Ficus hispida in China," Nematology, vol. 9, no. 2, pp. 169–187, 2007

106. E. Van Den Berg, S. A. Subbotin, Z. A. Handoo, and L. R. Tiedt, "Hirschmanniella kwazuna sp. n. from South Africa with notes on a new record of H. spinicaudata (Schuurmans Stekhoven, 1944) Luc & goodey, 1964 (nematoda: Pratylenchidae) and on

the molecular phylogeny of Hirschmanniella Luc & goodey, 1964," Nematology, vol. 11, no. 4, pp. 523–540, 2009.

107. C. D. Giorgi, P. Veronico, F. D. Luca, A. Natilla, C. Lanave, and G. Pesole, "Structural and evolutionary analysis of the ribosomal genes of the parasitic nematode Meloidogyne artiellia suggests its ancient origin," Molecular and Biochemical Parasitology, vol. 124, no. 1-2, pp. 91–94, 2002.

108. G. Anthoine and D. Mugniéry, "Variability of the ITS rDNA and identification of Nacobbus aberrans(Thorne, 1935) Thorne & Allen, 1944 (Nematoda: Pratylenchidae) by rDNA amplification," Nematology, vol. 7, no. 4, pp. 503–516, 2005.

109. L. Al-Banna, V. Williamson, and S. L. Gardner, "Phylogenetic analysis of nematodes of the genusPratylenchus using nuclear 26S rDNA," Molecular Phylogenetics and Evolution, vol. 7, no. 1, pp. 94–102, 1997.

110. X. Li and J. Zheng, "Identification of four Pratylenchus species based on morphology and PCR-RFLP of rDNA-ITS," Acta Phytopathologica Sinica, vol. 43, no. 4, pp. 444–448, 2013.

111. M. T. Britton, C. A. Leslie, G. H. McGranahan, and A. M. Dandekar, "Functional genomic analysis of walnut-nematode interactions," Walnut Research Reports 2009, California Walnut Board, 2010.

112. J. C. Wang, G. M. Huang, and Y. D. Wei, et al., "Phylogenetic analysis of Pratylenchus (Nematoda: Pratylenchidae based on ribosomal internal transcribed spacers (ITS) and D2/D3 expansion segments of 28S rRNA gene," Acta Zootaxonomica Sinica, vol. 4, pp. 687–693, 2012.

113. D. Y. Chen, H. F. Ni, and T. T. Tsay, "Identification of a new recorded stunt nematode Tylenchorhynchus zeae (Nematoda: Belonolaimidae) in Taiwan," Plant Pathology Bulletin, vol. 16, no. 2, pp. 79–86, 2007.

114. J. Liu, R. E. Berry, and A. F. Moldenke, "Phylogenetic relationships of entomopathogenic nematodes (Heterorhabditidae and Steinernematidae) inferred from partial 18S rRNA gene sequences," Journal of Invertebrate Pathology, vol. 69, no. 3, pp. 246–252, 1997

115. M. V. Blanco, P. Lax, J. C. Rondan Dueñas, C. N. Gardenal, and M. E. Douc, "Morphological and molecular characterization of

the entomoparasitic nematode Hammerschmidtiella diesingi (Nematoda, Oxyurida, Thelastomatidae)," Acta Parasitologica, vol. 57, no. 3, pp. 302–310, 2012.

116. S. A. Nadler, E. Bolotin, and S. P. Stock, "Phylogenetic relationships of Steinernema Travassos, 1927 (Nematoda: Cephalobina: Steinernematidae) based on nuclear, mitochondrial and morphological data,"Systematic Parasitology, vol. 63, no. 3, pp. 161–181, 2006

117. O. Douda, M. Zouhar, E. Nováková, J. Mazáková, and P. Ryšánek, "Variability of D2/D3 segment sequences of several populations and pathotypes of potato cyst nematodes (Globodera rostochiensis,Globodera pallida)," Plant Protection Science, vol. 46, no. 4, pp. 171–180, 2010.

118. S. P. Stock, J. F. Campbell, and S. A. Nadler, "Phylogeny of Steinernema travassos, 1927 (Cephalobina: Steinernematidae) inferred from ribosomal DNA sequences and morphological characters," Journal of Parasitology, vol. 87, no. 4, pp. 877–889, 2001.

119. E. E. Morris, R. M. Kepler, S. J. Long, D. W. Williams, and A. E. Hajek, "Phylogenetic analysis ofDeladenus nematodes parasitizing northeastern North American Sirex species," Journal of Invertebrate Pathology, vol. 113, pp. 177–183, 2013.

120. Q. Yu, P. de Groot, I. Leal, C. Davis, W. Ye, and B. Foord, "Characterization of Deladenus siricidicola(Tylenchida: Neotylenchidae) ssociated with Sirex noctilio (Hymenoptera: Siricidae) in Canada,"Intemational Joumal of Nematology, vol. 19, no. 1, pp. 23–32, 2009.

121. K. Zhang, H. Liu, and J. Sun et al., "Molecular phytogeny of geographical isolates of Bursaphelenchus xylophilus: implications on the origin and spread of this species in China and worldwide," Journal of Nematology, vol. 40, no. 2, pp. 127–137, 2008.

122. R. B. Gasser and H. Hoste, "Genetic markers for closely-related parasitic nematodes," Molecular and Cellular Probes, vol. 9, no. 5, pp. 315–319, 1995.

123. T. O. Powers, T. C. Todd, and A. M. Burnell et al., "The rDNA internal transcribed spacer region as a taxonomic marker for nematodes," Journal of Nematology, vol. 29, no. 4, pp. 441–450, 1997.

124. R. Floyd, E. Abebe, A. Papert, and M. Blaxter, "Molecular barcodes for soil nematode identification,"Molecular Ecology, vol. 11, no. 4, pp. 839–850, 2002.

125. A. Eyualem and M. Blaxter, "Comparison of biological, molecular, and morphological methods of species identification in a set of cultured Panagrolaimus isolates," Journal of Nematology, vol. 35, no. 1, pp. 119–128, 2003.

126. P. De Ley, I. T. De Ley, and K. Morris et al., "An integrated approach to fast and informative morphological vouchering of nematodes for applications in molecular barcoding," Philosophical Transactions of the Royal Society B, vol. 360, no. 1462, pp. 1945–1958, 2005.

127. N. R. R. Da Silva, M. C. Da Silva, and V. F. Genevois et al., "Marine nematode taxonomy in the age of DNA: the present and future of molecular tools to assess their biodiversity," Nematology, vol. 12, no. 5, pp. 661–672, 2010.

128. M. A. Smith, D. M. Wood, D. H. Janzen, W. Hallwachs, and P. D. N. Hebert, "DNA barcodes affirm that 16 species of apparently generalist tropical parasitoid flies (Diptera, Tachinidae) are not all generalists,"Proceedings of the National Academy of Sciences of the United States of America, vol. 104, no. 12, pp. 4967–4972, 2007.

129. H. van Megen, S. van den Elsen, and M. Holterman et al., "A phylogenetic tree of nematodes based on about 1200 full-length small subunit ribosomal DNA sequences," Nematology, vol. 11, no. 6, pp. 927–950, 2009.

130. A. Yu. Ryss, "Phylogeny of the order Tylenchida (Nematoda)," Russian Journal of Nematology, vol. 1, no. 2, pp. 74–95, 1993.

131. L. K. Carta, A. M. Skantar, and Z. A. Handoo, "Molecular rDNA phylogeny of telotylenchidae siddiqi, 1960 and evaluation of tail termini," Journal of Nematology, vol. 42, no. 4, pp. 359–369, 2010.

132. R. A. Bedding, "Controlling the pine-killing woodwasp, Sirex noctilio, with nematodes," in Use of Microbes For Control of Invasive Arthropods, A. E. Hajek, T. R. Glare, and M. O'Callaghan, Eds., pp. 213–235, Springer, Amsterdam, The Netherlands, 2009.

133. A. R. Maggenti, M. Luc, D. J. Raski, R. Fortuner, and E. Geraert, "A reappraisal of Tylenchina (Nemata). 2. Classification of

the suborder Tylenchina (Nemata: Diplogasteria)," Revue De Nématologie, vol. 10, no. 2, pp. 135–142, 1987.

134. V. N. Chizhov and S. N. Kruchina, "Phylogeny of the nematode order Tylenchida (Nematoda),"Zoologichesky Zhurnal, vol. 67, no. 9, pp. 1282–1293, 1988.

135. M. Luc, A. R. Maggenti, R. Fortuner, D. J. Raski, and E. Geraert, "A Reappraisal of Tylenchina (Nemata) 1. For a New Approach to the Taxonomy of Tylenchina," Revue De Nématologie, vol. 10, no. 2, pp. 127–134, 1987.

136. O. V. Slobodyanyuk, "Host specificity in tylenchids of fleas," in Abstracts of Papers Presented at the Russian Society of Nematologists 1st English Language International Symposium, pp. 102–103, Zoological Institute of the Russian Academy of Sciences, 1995.

137. K. D. Elsey, "Parasitism of some economically important species of chrysomelidae by nematodes of the genus howardula," Journal of Invertebrate Pathology, vol. 29, no. 3, pp. 384–385, 1977.

138. G. O. Poinar Jr., J. Jaenike, and D. D. Shoemaker, "Howardula neocosmis sp.n. parasitizing North American Drosophila (Diptera: Drosophilidae) with a listing of the species of Howardula Cobb, 1921 (Tylenchida: Allantonematidae)," Fundamental and Applied Nematology, vol. 21, no. 5, pp. 547–552, 1998.

139. R. L. H. Dennis, L. Dapporto, S. Fattorini, and L. M. Cook, "The generalism-specialism debate: the role of generalists in the life and death of species," Biological Journal of the Linnean Society, vol. 104, pp. 725–737, 2011.

140. C. E. Richmond, D. L. Breitburg, and K. A. Rose, "The role of environmental generalist species in ecosystem function," Ecological Modelling, vol. 188, no. 2-4, pp. 279–295, 2005.

141. M. E. J. Woolhouse, L. H. Taylor, and D. T. Haydon, "Population biology of multihost pathogens,"Science, vol. 292, no. 5519, pp. 1109–1112, 2001

142. H. D. Loxdale, G. Lushai, and J. A. Harvey, "The evolutionary improbability of "generalism" in nature, with special reference to insects," Biological Journal of the Linnean Society, vol. 103, no. 1, pp. 1–18, 2011.

143. J. C. Castillo, S. E. Reynolds, and I. Eleftherianos, "Insect immune responses to nematode parasites,"Trends in Parasitology, vol. 27, no. 12, pp. 537–547, 2011.

144. R. Kassen, "The experimental evolution of specialists, generalists, and the maintenance of diversity,"Journal of Evolutionary Biology, vol. 15, no. 2, pp. 173–190, 2002.

145. N. I. Samia, K. L. Kausrud, and H. Heesterbeek et al., "Dynamics of the plague-wildlife-human system in Central Asia are controlled by two epidemiological thresholds," Proceedings of the National Academy of Sciences of the United States of America, vol. 108, no. 35, pp. 14527–14532, 2011.

146. H. E. Kaiser and A. J. Boucot, "Specialisation and extinction: cope›s law revisited," Historical Biology, vol. 11, no. 1–4, pp. 247–265, 1996.

147. C. Darby, A. Chakraborti, S. M. Politz, C. C. Daniels, L. Tan, and K. Drace, "Caenorhabditis elegansmutants resistant to attachment of Yersinia biofilms," Genetics, vol. 176, no. 1, pp. 221–230, 2007.

148. G. A. Eroshenko, N. A. Vidyaeva, and V. V. Kutyrev, "Comparative analysis of biofilm formation by main and nonmain subspecies Yersinia pestis strains," FEMS Immunology and Medical Microbiology, vol. 59, no. 3, pp. 513–520, 2010.

149. G. A. Eroshenko, N. A. Vidyaeva, and L. M. Kukleva, et al., "Studies of biofilm formation in non-pigmented and plasmid-deprived mutants of Yersinia pestis on biotic surfaces, in vivo and in vitro conditions,"Problems of Particularliy Dangerous Infections, vol. 113, pp. 45–49, 2012.

chapter 2

Commercial Biological Control Agents Targeted Against Plant-Parasitic Root-knot Nematodes

Marie-Stéphane Tranier[1], Johan Pognant-Gros[1], Reynaldo De la Cruz Quiroz[2], Cristóbal Noé Aguilar González[2], Thierry Mateille[3], and Sevastianos.Roussos[1]

[1]Institut Méditerranéen de Biodiversité et d'Ecologie marine et continentale (IMBE); UMR CNRS IRD; Aix Marseille Université; Marseille - France

[2]Departamento de Investigación en Alimentos; Universidad Autónoma de Coahuila; Saltillo - México

[3]Centre de Biologie pour la Gestion des Populations (CBGP); IRD; Montferrier sur Lez Cedex - France

ABSTRACT

Root-knot nematodes are microscopic round worms, which cause severe agricultural losses. Their attacks affect the productivity by reducing the amount and the caliber of the fruits. Chemical control is widely used, but biological control appears to be a better solution, mainly using microorganisms to reduce the quantity of pests infecting crops. Biological control is developing gradually, and with time, more products are being marketed worldwide. They can be formulated with bacteria, viruses or with filamentous fungi, which can destroy and feed on phytoparasitic nematodes. To be used by the farmers, biopesticides must be legalized by the states, which has led to the establishment of a legal framework for their use, devised by various governmental organizations.

INTRODUCTION

Root-knot nematodes are disease-causative agents, well-known by farmers, for example, tomato producers in Morocco. Each year, these microscopic worms cause considerable losses. Biopesticides may be a good solution to reduce the damages, but they have to be legalized by the governments. Filamentous fungi may play a good role in the control of this pest, using different action pathways. They have to be formulated to preserve their viability and virulence against the nematodes, and then to be approved for use on the crops by farmers.

ROOT-KNOT NEMATODES (RKN)

Nematodes are classified under the *Nematoda* Phylum, which forms part of the *Ecdysozoa* superphylum, from the Greek ecdysis (moulting) and zoo (animal). They possess a cuticle and must moult to continue their development (Bélair 2005). Most of them are said "free" and feed essentially on bacteria, fungi, protozoa and other nematodes, and only a minority parasites animals and plants (Bélair 2005).

Meloidogyne spp. or RKN are round worms (Brand et al. 2009) belonging to *Tylenchida* family. They are cylindrical and thread-like triploblastic Metazoans that show bilateral symmetry, a thick cuticle, and non-ciliated, non-segmented teguments (Grassé et al. 1965). They were discovered during the 14th century as phytoparasites of intertropical and warm climate cultures. The agricultural losses they cause, in particular in greenhouse cultures of tomatoes in the Souss-Maasra Drâa region in Morocco, are considerable. Worldwide, more than USD 125 billion/year are lost due to nematodes, most of them due to RKN (Koenning et al. 1999; Chitwood 2003; Collange et al. 2011; Bissadou et al. 2012). The degree of these damages can be even more severe as nematodes can increase susceptibility to other parasites (Bridge et al. 2005).

The *Meloidogyne* genus subdivides in different species, all phytophagous, with *M. arenaria*, *M. incognita*, *M. hapla* and *M. javanica* being commonly widespread (Netscher 1970; Sawadogo et al. 2000; Hunt and Handoo 2009). The number of plants susceptible to *Meloidogyne* spp. is very high (more than 2,000 species, a constantly increasing figure), and includes tobacco (*Nicotiana tabacum*), coffee (*Coffea arabica*), tomato (*Solanum lycopersicum),*Niebe (*Vigna sinensis*), Kenaf (*Hibiscus cannabinus L.)*, sugar cane (*Saccharum officinarum*), tea (*Camellia sinensi*), carrot (*Daucus carota*), melon (*Cucumis melo*), etc. *Meloidogyne* spp are round worms, 0.4 to 1.0 mm long and 0.25-0.75 μm in diameter for the females and 1.2-1.5 mm long and 30-36 μm in diameter for the males (Agrios 1997). These present a stylet, which is used to pierce the cell walls of the roots of the host plants (Davis et al. 2000). As they enter the root, the larvae will provoke a cortical cell hypertrophy due to the excretion of secretions through the stylet. Multi-nucleated «giant cells» are thereby formed (Bird 1962), which cause the formation of gall, which is characteristic of the attack by a *Meloidogyne* spp.

Once hatched, J2 free larvae become obligatory parasites that have to continue their cycle into adult stage in host-plant roots only (Sharon et al. 2001). For that purpose, they swim into the water pellicle surrounding the soil particles. Once they penetrate the root thanks to the stylet, the larvae move both in and between the cells. They move to the central cylinder along which they immobilize (Fig. 1A). At this time, L3 and L4 larvae turn into sexual adults, presenting strong dimorphism (Fig. 1B): males stay vermiform whereas females become pyriform and sedentary, their heads lodged in giant cells. Once fecundated, females lay eggs (around 500/mass) agglomerated at all the stages of development in a gelatinous substance, from the unicellular stage to the ready-to-hatch stage. Egg development between these two stages could take from seven to nine days at 28°C. During this time, nematodes undergo a first molting and the larvae that hatch are J2 second-stage larvae (De Guiran and Netscher 1970).

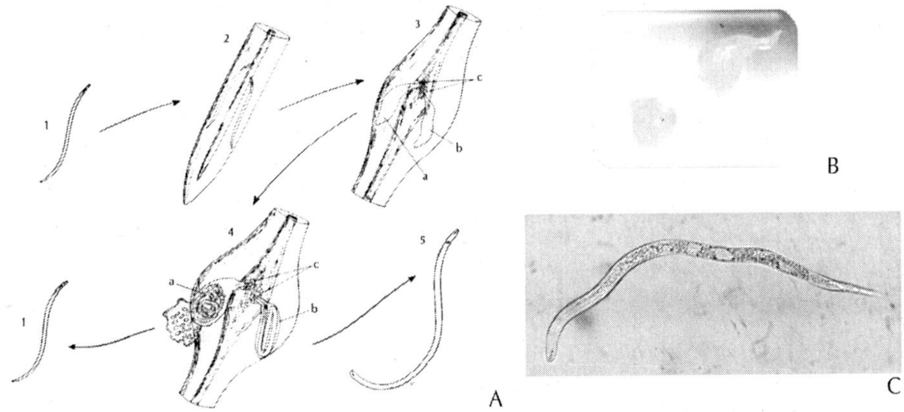

Figure 1: A. Schematized cycle of Meloidogyne spp. - 1. Free second stage larva; 2. Second stage larva penetrating a growing root; 3. Root in formation : a-b : Swolled second stage larva ; c : Giant cells ; 4. Gall containing : a : adult female which had laid her eggs in the gelatinous substance ; b : adult male curled up in the larval envelopes ; c : Giant cells; 5. Free adult male (By De Guiran G. and Netscher C. 1970) ; B. Photo of females of M. javanica (M. Tranier) and C. J2 larvae of M. javanica (M. Tranier).

A higher temperature may accelerate the cycle; yet temperatures above 40°C are lethal. Eggs are the nematodes' resistant form, so they survive unfavorable conditions of life. In the absence of males, females are also able to produce fertile eggs in a parthenogenetic way (Agrios 1997). Typical symptoms of damage caused by the nematodes are a reduction of the root system, root structure distortion or root diameter increase, and gall presence. Plants turn yellow and wither and a production decrease is observed (Agrios 1997).

BIOLOGICAL CONTROL

General Definition

Founded in 1956, the International Organization of Biological Control (Iobc 2014) is set to promote environmentally-, socially- and-economically-efficient methods to eliminate the diseases and pests from agriculture and forests. It defines biological control as "the use of living organisms or their products to eliminate or reduce the damages or losses due to pests ". Biological control involves the use of beneficial organisms, of their genes and/or of products such as metabolites that reduce the negative effects of plant pathogens and promote positive responses from the plant (Junaid et al. 2013). More precisely, it uses living populations of parasitoid, predatory, pathogenic, antagonist or competing organisms to reduce a pest population, making it less abundant and less harmful than it would otherwise have been (De Bach 1964; Stirling 1991; Siddiqui and Mahmood 1996). The Biological Control Agent (BCA) is then the organism or its product which combats the pest. It is usually called a "biopesticide ".

Biopesticides are generally defined as the "products aimed at protecting the plants, made from living organisms or natural substances from species co-evolution, not produced by chemistry, and the use of which is recommended for the control of pests or bio-aggressors for a better respect of the biocenosis and environment". According to this definition, Genetically Modified Organisms (GMO) that are not issued from the co-evolution of species and which have known too little anteriority in terms of impact on biological communities and the environment should so far be excluded from the field of biopesticides.

If the antagonist is a microorganism, microbiological control could be considered. In general, the microorganism infects the host by ingestion and has a resistant form that permits it to pass through - and stay in - the medium (soil, foliage, litter). The pathogen multiplies in the host and causes its death by tissues destruction, septicemia, and sometimes the emission of a toxic substance (case of bacteria). Host cadavers release the pathogenic agents in the medium. When the antagonists have to be inoculated in large numbers to effectively reduce the pest, the method is called inundative biological control; it is frequently used, but poses the problem of the future of the auxiliary populations introduced in the field.

Biological control appears to be an excellent alternative to the chemical one, which is the major method used to fight the pests. It consists in a general disinfection of fields (before each plantation) with fumigants or precursors, but presents a real danger for the environment and human health. These chemical products indeed present serious drawbacks (wide action spectra, disturbed ecological balances, pollution of the environment and food-chain, insufficient results, long-term resistance of pests). This method of control also poses the problem of the necessary delay between the product's application and harvest. Moreover, widely-used chemical products (like methyl bromide) are increasingly banned, which poses the problem of the existence of alternatives, if possible within the legal framework of organic farming. However some alternatives to chemical products do exist, which are used but present limited effects: inundation and summer solarization are two examples (Van der Putten et al. 2006; Djian-Caporalino et al. 2009; Collange et al. 2011). Inundation causes anaerobic conditions that reduce the occurrence of RKN, but it requires a long flooding period (eight weeks) to be efficient, and causes water consumption, lack of oxygen and degradation of the soil's structure, which is not suitable to agronomic practices (Collange et al. 2011). Summer solarization, in the intercropping period, consists in covering the soil with a plastic film during the hottest months of the year to increase the soil's temperature. It helps neutralize thermosensible pathogens at depths up to 20 cm, like root-knot nematodes. However, the heat can't penetrate into the deeper layers of the soil where nematodes can also shelter. Soil amendments with composts or manures can stimulate natural microbial soil flora and have a repressive effect on phytoparasitic nematode populations (Bélair 2005). Over the last few years, biological control has, thus, been developing. This article presents a review on nematode-antagonistic fungi and some of the commercial products associated.

FUNGI AS BCA AGAINST MELOIDOGYNE SPP

Fungi naturally exist that have an antagonistic action against the nematodes in general (Sarhy-Bagnon et al. 2000, Brand et al. 2004), more precisely *Meloidogyne* spp. (Viaene and Abawi, 1998a; Stirling et al. 1998; Stirling and Smith 1998; Duponnois et al. 1998; 2001; Kumar and Singh 2006; Thakur and Devi 2007; Liu et al. 2008; Collange et al. 2011). The infection process is generally the same: cuticle penetration, immobilization, invasion and digestion of the nematode (Huang et al. 2004). The nematodes' cuticles and egg walls play an important role in the infestation by fungi. The cuticle is essentially made of proteins (chitin, collagen, fibers) and can play the role of a precursor in the invasion of the nematodes by nematophagous fungi (Huang et al. 2004). Whatever the mode of action of the fungus, predation or parasitism, the interaction between a fungus and a host organism or prey requires penetration of the outer shell of the body. This penetration occurs before the colonization of the internal structures of the body through digestion, allowing the fungus to meet its nutritional needs (Gaspard et al. 1990;Hajieghrari et al. 2008). Cayrol et al. (1992) classified the nematophagous fungi according to their mode of action against the nematodes.

Predatory Fungi (Cayrol et al. 1992)

Some fungi have the ability to trap the nematodes they feed on. They differ by their trapping mode: network traps (*Arthrobrotrys oligospora, A. superba*), constrictive rings (*A. anchonia, A. dactyloides, Dactylaria brochopaga, etc*), or adhesive knobs (*Monacrosporium cionopagum, Dactylella lobata*). Some fungi, such as*Dactylaria candida* (Hyphomycete) present two types of trapping mechanism: i) adhesive knobs, and ii) constrictive but non-adhesive rings (B'Chir 1984). These fungi are saprophytes, but they can trap the nematodes in a larval or free adult stage in the soil. It has been proven that the production of peptides by extracellular proteases hydrolyzing the nematodes' cuticles may induce trap formation (Huang et al. 2004).

Egg-Parasitic Fungi (Cayrol et al. 1992)

These fungi have the ability to infect and destroy nematode's eggs. Most of the fungi that have this mode of action are saprophytes, and can secondarily invade already-dead eggs. Egg-parasitic fungi include*Paecilomyces*, *Pochonia* and *Verticilium* genera. *Paecilomyces lilacinus* and *Pochonia chlamydosporia* are probably the most effective egg-parasites. *P. lilacinus* has been proven to successfully control root-knot nematodes, *M. javanica* and *M. incognita* on tomato, egg-plant and other vegetable crops (Cayrol et al. 1989;Verdejo-Lucas et al. 2003; Goswami and Mittal 2004; Van Damme et al. 2005; Goswami et al. 2006; Haseeb and Kumar 2006).

Nematophagous Fungi with Adhesive Spores (Cayrol et al. 1992)

Nematodes can be parasited by the fungi with adhesive spores belonging to several classes: Oomycetes (*Catenaria anguillulae*, *Myzocytium lenticulare*, *M. anomalum* with biflagellated zoospores able to encyst in the nematode's cuticle), Zygomycetes (*Meristracum asterospermum*, with spherical conidia, which produce a germinative filament after adhesion that germinates and creates new conidiospores), Deuteromycetes (*Meria coniospora* with club-like spores, which fix on the host via their anterior part), and Basidiomycetes (*Nematoctonus leiosporus*, with adhesive spores) and Hyphomycetes (*Hirsutella sp*). Exclusively endoparasitic fungi must live their entire cycle in the host nematodes.

Endomycorhizes with Arbuscular Vesicles (Cayrol 1992)

Mycorhizes are fungi associated with the plant roots in a symbiotic way, presenting a double mycelial network: one external, in the soil, and one internal, in the tissues of the host-plant. The mycelial network expands widely in the rhizosphera and exploits a large volume of soil, which ensures abundant nutrition for host-roots, increasing the assimilative potential of the plant. Nematodes living in these roots compete in intra-root nutrition and face a plant becoming more resistant, thanks to the nutritional amelioration due to mycorhizes.

Microbial Toxins

Some substances are naturally nematifuge (repulsives), nematostatic (perturbating host-plant reconnaissance by the nematode; blocking the development of the egg or the larva; or temporarily paralyzing the animal) or nematicide (*sensus stricto*: ovicide, larvicide or lethal for all stages). Some studies have been conducted, but it seemed that some enzymes produced by the fungi, or volatile organic compounds (e.g., 6-pentyl-α-pyrone, produced by *Trichoderma harzianum*), could have some antagonistic effects on nematodes (Sarhy-Bagnon et al. 2000). It has been proven that the toxin produced by *Bacillus thuringiensis* is efficient *in vitro* (Cayrol et al. 1992)

Actinomycetes against Meloidogyne

Gram-positive parasite bacteria such as actinomycetes (e.g. *Pasteuria penetrans)* possess BCA qualities against phytoparasitic nematodes (Mateille 1993). Spores staying in soil adhere to the J2 when they pass near the colonies. Adhesion is very specific of the *Meloidogyne* genera and of its larval stage. Spore germination occurs eight days after the J2 penetrates the root and starts feeding on it. This Actinomycete is perfectly synchronized with the development and the physiology of *Meloidogyne*. Then, a germinative tube is formed by the cuticle and infestation occurs. At the end, *Pasteuria penetrans* forms new hypha and spores inside the females, which become sterile as their reproductive system is destroyed.

Other Organisms

There are carnivorous predatory nematodes belonging essentially to the order of *Mononchida*, to the *Diplogasteridae* suckers family, and to the *Dorylaimidae* omnivorous family, and others with toxic effects (Aphelenchidae). Some insects can also be nematophagous such as *Entomobryoides dissimilis* or nematodes' predatory mites (*Alliphis* spp., *Alicorhagia* spp.). But if the predation of these organisms is indisputable, their effects on nematode populations are unknown.

MAIN AGENCIES FOR THE HOMOLOGATION OF BCA'S, AND THEIR MARKET

Biocontrol agents and biopesticides must undergo an approval process before being placed on the market. This is a long and costly process but makes sure that the product brought to the field is both safe for humans and the environment. There are government organizations (agencies) responsible for defining the rules for using these products that come into contact with the food chain.

THE ENVIRONMENTAL PROTECTION AGENCY (EPA)

This U.S. government agency (United States 2014) is responsible for assessing the risks of the products used in the crops for food. Its mission is to protect both human health and the environment. When the Congress passes an environmental law, the EPA writes the regulations that strengthen it. This is the organization that certifies "biopesticide" products in the U.S., a leader in the field of biopesticides. In 2008, 279 biopesticides were registered in the U.S. compared to 77 in Europe, due to a procedure which is more complex and time consuming for the latter. However, the EPA does not require testing efficiency in its reviewing process. One can thus register a biopesticide on the sole basis of its safety for human health and the environment without demonstrating its effectiveness under different conditions for which the product may have to be used (Caron et al. 2006).

The Pest Management Regulatory Agency (ARLA)

The ARLA (Santé - Canada 2014) of Canada is committed to promoting, verifying and ensuring the compliance with the Act on the label. This is equivalent to the U.S. EPA. Unlike the EPA, the ARLA requires that the pesticide potential of a product be demonstrated, and that, over several seasons and over all particular geographical areas of Canada. This requirement leads to huge additional costs for the companies, which often cause the product not to meet the market's expectation. However, authorization by ARLA can be granted on the sole basis of EPA's registration of the same product (Caron et al. 2006).

The French Ministry of Agriculture and Fishing

It is responsible for assigning the AMM (Marketing Authorization) after assessing the risks of the product under investigation. For this, it relies on the Office of Pesticides and Fertilizers. Risk assessment is carried out by a committee of experts. The work of the experts and of the Commission for the Study of Toxicity is coordinated by the SSM (Joint Scientific Structure). The Accreditation Committee then makes a proposal, based on risk and efficiency assessments. The AMM, preceded by a provisional marketing authorization, is issued by the State and is valid throughout the European territory (France 2014).

THE ORGANIC MATERIALS REVIEW INSTITUTE (OMRI)

Founded in 1997, the Organic Materials Review Institute (OMRI) is a nonprofit organization (Omri 2014), which provides organic certifiers, growers, manufacturers, and suppliers an independent review of products intended for use in certified organic production, handling, and processing. When companies apply, OMRI reviews their products against the National Organic Standards. Acceptable products are OMRI-listed(r) and appear on the *OMRI Products List*. OMRI also provides subscribers and certifiers guidance on the acceptability of various material inputs, in general under the National Organic Program.

The Market for Biological Control Agents

The global pesticide market is constantly changing and was estimated at USD 43 billion in 2009, with predictions for an annual increase of 5 to 8% for each subsequent year. Two competitors, chemical pesticides and biopesticides, share this market. With recent developments in global market requirements for pesticide residues, and the desire for products from organic agriculture, biopesticides have gained much attention. Exponential growth is predicted in this sector, which currently accounts for USD 1.6 billion. The United States are a leader in this area but the most important market growth for biopesticides is in Europe. The market in Europe was USD 270 million in 2010 and a 15% increase was expected in this sector. France ranked third on this market, behind Spain and Italy (Source: Europe: Pesticides Biomarket, CPL Consultants 2010). In Europe, several companies compete in this market: Novartis (18.4%), Koppert (18.2%), Abbot Laboratories (12.5%), AgrEvo Intrachem, BASF, Bayer cropscience, etc.

MAJOR FUNGAL PREPARATIONS ACTIVE ON MELOIDOGYNE AVAILABLE ON THE MARKET

In the U.S., more than 279 biopesticides (all inclusive) were approved in 2009 against only 77 in France. This was due to a more restrictive policy on the part of the European Union. Fourteen bacterial and twelve fungi have been registered with the EPA for the control of plant diseases (Fravel 2005). Among the products available on the market, only a few of the best-known biological nematicides are given below.

BIO-ACT(r)/ Melocon Wg (Prophyta 2014)

Marketed by Prophyta (acquired in 2013 by Bayer CropSscience Biologics GmbH), BioAct(r) is a biological nematicide used for biocontrol of the root-knot nematodes (Brand et al. 2009). It contains 1010 spores per dry weight gram of the ovicidal fungus, *Paecilomyces lilacinus* strain 251, which is a highly effective parasite of all the stages of development of common plant-infecting nematodes, especially the eggs and infectious juveniles. Formulated as a water-dispersible granule, BioAct(r)/MeloCon can be applied through conventional methods, using the irrigation system. *Paecilomyces lilacinus* is filamentous Ascomycete fungus that controls phytoparasitic nematodes such as *Meloidogyne* spp. (Kiewnick and Sikora 2004; Kiewnick and Sikora 2006; Khan et al. 2006; Mukhtar et al. 2013), *Radopholus similis* (Mendoza et al.

2007), *Heterodera* spp., *Globodera* spp. and*Pratylenchus* spp. Direct interactions between *P. lilacinus* strain 251 and eggs but also females of *M. javanica*were demonstrated *in vitro* by Holland et al. (1999).

Botanigard(r); Mycotrol(r)

In 2009, the ARLA, under the authority of the Pest Control Products Act and Regulations, made a full registration of the Technical *Beauveria bassiana*, named Botanigard(r) ES or Botanigard(r)22WP (Laverlam 2014) and Mycotrol O(r) or Mycotrol ES(r) (Laverlam 2014), containing the *B. bassiana* strain GHA, a technical-grade active ingredient for sale and use in against whiteflies, aphids and thrips in vegetable and ornamental greenhouses. These products are also registered by the EPA. The *B. bassiana* GHA is found naturally in the soils throughout the world and parasitizes various insect species. It is a "generalist entomopathogenic fungus" - a fungus that causes disease in many types of insects. As the insects that live in the ground or near the ground have evolved natural defenses against *B. bassiana* because it is common in the wild, this fungus can be used as a biological insecticide against most other insects. Although its approval does not stipulate it, *B. bassiana* has a repressive action on nematodes of the genus *Meloidogyne* spp (Bradley et al. 1992; Bekanayake and Jayasundar 1994; Liu et al.2008).

BIOSTAT(r)

BIOSTAT(r) (EPA approved) contains a biotype of *Purpureocillium lilacinum*, selected for its capacity to control plant pathogenic nematodes, such as: *Meloidogyne* sp., *Radopholus* sp., *Pratylenchus* sp., *Tylenchus* sp.,*Ditylenchus* sp., *Helicotylenchus* sp. and *Rotylenchulus* sp. It parasitizes all reproductive stages of plant pathogenic nematodes, especially eggs and females, causing deformations, destruction of ovaries and reducing egg fertility. Also, under acid pH conditions, it produces toxins that affect the nematode's nervous system (Laverlam 2014).

MYCOTAL(r) (Koppert 2014)

Developed by KOPPERT, Mycotal(r) contains an entomopathogenic fungus, *Verticillium lecanii*. It is formulated as a wettable powder. It also acts against the larval stages of whiteflies, with a strong secondary action on adult thrips and red spiders. It has an action on the suppression on *M. incognita* on tomatoes, as reported by Meyer (1999).

TRIANUM(r) (Koppert 2014)

The use of *Trichoderma* spp. as biological control agents has been investigated for over 70 years but it is only relatively recently that strains have become commercially available. Many *Trichoderma* strains, mainly *T. harzianum*, *T. viride* (Masadeh et al. 2004) and *T. virens* (formerly *Gliocladium virens*) have been identified as having potential applications in biological control (Lumsden et al. 1993; Monte and Llobell 2001). Trianum(r) is a product marketed by KOPPERT based on *T. harzianum* T-22. This filamentous fungus has various advantages for the protection of plants and the fight against nematodes (Widden and Abitbol 1980; González et al. 2012): it produces molecules such as 6-pentyl α-pyrone, VOCs and enzymes (Samson et al. 1996) that can attack the cuticle of nematodes. Also, its hyphae form a physical barrier, which is a difficult step for nematodes, since the fungus grows along with the plant roots. *T. harzianum* has a beneficial effect on the stimulation of the defense system of the plants (Samuels 1996; Harman 2000).

MET52(r) (Novozymes 2014)

Metarhizium anisopliae is an entomopathogenic fungus used against thrips, termites and mosquitoes. Studies have shown its nematicide potential as well (Mahmoud 2009). The PMRA-ARLA granted conditional approval in 2010 for the sale and use of *M. anisopliae* strain F52 and the associated end-bio-Met52 Granular Insecticide. Its active ingredient is the F52 strain of *M. anisopliae* in order to remove the root weevils, particularly black vine weevils and strawberry root weevils, which infest ornamental plants grown in pots.

NEMAXXION BIOL(r) (Greencorp 2014)

Nemaxxion Biol(r) has been commercialized by GreenCorp (Mexico). It is a liquid-formulated large-spectra product that is active against nematodes, composed of a consortium of microorganisms (*Bacillus subtilis,Trichoderma* spp, *Paecilomyces* spp.) and extracts (tagetes) that are active against RKN.

REM G(r)

REM G(r) is a new product commercialized by Green Solutions in Italy. Formulated as a consortium of nematophagous fungi such as *Arthrobotrys* spp, *Dactyllela* spp., *Paecilomyces* spp., Mycorrhiza (*Glomus* spp.), and bacteria (*B.* spp. and *Pseudomonas* spp.), it is supplemented with chitinolytic, proteolytic and lipolytic enzymes to specifically target the nematodes' walls. It is currently showing good results in tomato crops in the Souss-Massa Drâa aera in Morocco (personal information).

Production of BCAs

There are two ways of producing filamentous fungi (or bacteria, such as *Pasteuria penetrans*), by Liquid-State Fermentation (LSF) or Solid-State Fermentation (SSF). The choice of the method depends on the desired (liquid or wettable powder) final formulation. It also depends on the strain and on production costs.

PRODUCTION BY LIQUID-STATE FERMENTATION

Fermentation in a liquid medium is conducted in large, agitated, temperature-controlled and aerated tanks, and is perfectly suited to the production of single-celled organisms such as bacteria or yeasts. Some fungi can be produced by LSF such as *Fusarium venenatum*, used as Qorn(r), for human food. Examples of microorganisms produced by LSF include *P. penetrans, B. thuringiensis,* etc.

PRODUCTION BY SOLID-STATE FERMENTATION

SSF is generally known as a culture of microorganisms (preferably filamentous fungi) on a solid substrate without water flow (Hesseltine 1987; Roussos et al. 1993; Mitchell 2002; Barrios-González 2012). It exploits the metabolism and growth mechanism of these microorganisms to degrade the solid substrates to produce biopesticides (biomass and secondary metabolites). Microbial growth occurs on the surface and within the solid matrix in the absence of any liquid flow. The porous matrix may be formed from a natural substrate or from an inert support capable of absorbing the nutrients found in dissolved state in a solution. For their mass multiplication, Selvakumar and Srivastava (2000) and Zaidi and Singh (2004), multiplied *T. harzianum* on pre-soaked and autoclaved Jhangora seeds for 12 days at 28°C, then air-dried, ground and passed them through 50 and 80 mesh sieves simultaneously to obtain spore powder. The commercial formulation was prepared by diluting this powder with talcum powder containing 1% carboxymethyl-cellulose to get the desirable concentration of biocontrol agent.

Formulation of BCAs

Formulation is the blending of active ingredients, such as fungal spores, with an inert material, such as diluents or surfactants, in order to preserve the viability and virulence of the strain used. Commercial products also have to present an appropriate structure (powder or liquid) for their application in the field. Formulation must ensure the viability of the strain, preserving its germinating power; it must help the strain keep its virulence against the pest involved. The material can be mineral such as talc (Chaube et al. 2003) or zeolites (Küçük and Kivanç 2005). It is also essential to protect the BCA from the UV rays of the sun and to stabilize the product for both storage and utilization conditions. A granular starch matrix encapsulating *B. thuringiensis* was patented in 1997 (Quimby et al. 1996). One of the technologies for the formulation of biocontrol organisms is the immobilization of wet or dry biomass within cross-linked polymers such as alginate and carrageen (Cho and Lee 1999). Incorporating fungal mycelia in alginate pellets has been found to be successful for the delivery of biocontrol fungi (Papavizas et al. 1987; Küçük and Kivanç 2005). The biocontrol microorganisms were immobilized, wet or dry, as formulated pellets (Walker and Connick 1983). Alginate-type pellets were used in the formulations of chemical and microbial herbicides (Walker and Connick 1983). In the biotechnology industry, cell entrapment is often used to enhance the production rates of bioproducts, to reduce the mortality of cells, and to facilitate their recovery. Alginate pellets containing the spores of various biocontrol fungi (Lewis and Papavizas 1983; 1985) and of yeast cells have been formulated (Serp et al. 2000). Such preparations offer many advantages compared with the conidial suspensions, e.g., pellets can be stored dry.

CONCLUSIONS

Biological control is a sustainable alternative to chemical control. The use of fungi to control the pests like root-knot nematodes may be developed to protect the environment from the pesticides. Unless they are more difficult to produce, store and/or use, they can be a suitable solution to protect the crops. Currently, some products are available on the market and more are being developed. Isolating indigenous strains is the best way to ensure the success of the use of these biological control agents, because they are then adapted to the pest they have to control and to the environment they will be used in. Formulation of these strains should be adapted to ensure both good conservation of the microorganisms and high virulence against the pests.

ACKNOWLEDGMENTS

This work was supported by IDYL SAS Company and AIRD scientific collaboration (EOTP N°4724A1).

REFERENCES

1. Agrios GN. Plant Pathology. Academic Press, fourth edition. San Diego USA. 1997

2. Barrios-González J. Solid-state fermentation: Physiology of solid medium, its molecular basis and applications. Process Biochem. 2012; 47: 75-185.

3. B'Chir MM. Étude de l›action des champignons prédateurs sur divers nématodes du sol en microscopie électronique à balayage (SEM). Revue Nematol. 1984; 7(1): 29-34.

4. Bekanayake HMRK, Jayasundar NJ. Effect of Paecilomyces lilacinus and Beauveria bassiana in controlling Meloidogyne incognita on tomato in Sri Lanka. Nematol Medit. 1994; 22: 87-88.

5. Bélair G. Les nématodes, ces anguillules qui font sur les plantes... par la racine. Phytoprotection. 2005; 86: 65-69.

6. Bird AF. Orientation of the larvae of Meloidogyne javanica relative to roots. Nematol. 1962; 8: 275-287.

7. Bissadou KD, Tchabi A, Tounou AK, Ayessom A, Gumedzoe M. Impact de la fumure organique appliquée seule et en combinaison avec une souche indigène de champignon mycorhizien arbusculaire Glomus mosseae sur Meloidogyne spp, principal nématode parasitaire de la tomate au Togo. J Appl Biosc. 2012; 55: 3973-3986.

8. Bradley CA, Black WE, Kearns R, Wood P. Role of production technology in mycoinsecticide development, pp. 160-173. In G. E. Leatham (ed.), Frontiers in industrial microbiology. Chapman & Hall, New York. 1992

9. Brand D, Oishi BO, Roussos S, Soccol CR. Spore production of Paecilomyces lilacinus by Solid State Culture and application in Pot Experiments to control Meloidogyne incognita. Appl Biochem Biotech. 2004; 118: 81-88.

10. Brand D, Soccol CR, Sabu A, Roussos S. Production of fungal biological control agents through solid state fermentation: a case study on Paecilomyces lilacinus against root-knot nematodes. Micol Appl Int. 2009; 21 (2): 33-50.

11. Bridge J, Plowright RA, Peng D. Nematode parasites of rice. In Plant parasitic nematodes in subtropical and tropical agriculture. 2ond edition. Edited by Luc M., Sikora R.A. and Bridge J. CABI Publishing. 2005.

12. Caron J, Laverdière L, Venne J, Bélanger R. Recherche et développement de biopesticides et pesticides naturels à faible

toxicité pour les organismes non ciblés et respectueux de l›environnement - Rapport final - Volet Phytopathologie. Projet PARDE. 278p. 2006.

13. Cayrol JC, Dijan C, Pijarowski L. Study of the nematocidal properties of the culture filtrate of the nematophagous fungus Paecilomyces lilacinus. Rev Nematol. 1989; 12 (4): 331-336.
14. Cayrol JC, Dijan-Caporalino C, Panchaud-Mattei E. La lutte biologique contre les Nématodes phytoparasites. Courrier de la cellule Environnement de l›INRA. 1992; 17: 31-44.
15. Cayrol JC, Djian C, Frankowski JP. Efficacy of Abamectin B1 for the control of Meloidogyne arenaria. Fund Appl Nematol. 1993; 16: 239-246.
16. Cayrol JC, Propriétés nématicides des endomycorhizes à vésicules et arbuscules. PHM- Rev Hort. 1991; 321: 33-42.
17. Chaube HS, Mishra S, Varshney DS, Singh SU. Biocontrol of plant pathogens by fungal antagonists: Historical background, present status and future prospects. Ann Rev Pl Pathol. 2003; 2: 1-42.
18. Chitwood DJ. Research on plant-parasitic nematode biology conducted by the United States Department of Agriculture-Agricultural Research Service. Pest Manag Sci. 2003; 59: 748-753.
19. Cho CC, Lee WC. Formulation of biocontrol agent by entrapping biomass ofTrichoderma viride in gluten matrix. J Biosci Bioengen. 1999; 87 (6) 822-824.
20. Collange B, Navarrete M, Peyre G, Mateille T, Tchamitchian M. Root-knot nematode (Meloidogyne) management in vegetable crop production: The challenge of an agronomic system analysis. Crop Prot. 2011; 30 (10): 1251-1262.
21. Davis EL, Hussey R.S, Baum TJ, Bakker J, Schots A, Rosso MN, Abad P. Nematode parasitism genes. Annu Rev Phytopathol. 2000; 38: 365-396.
22. De Guiran G, Netshter G. Les nématodes du genre Meloidogyne, parasites de cultures tropicales. Cahiers de l›ORSTOM. 1970; 11: 151-185.
23. DeBach P. Biological control of insect pests and weeds. New York: Reihold. 1964.

24. Djian-Caporalino C, Védie H, Arrufat A. De nouvelles pistes pour gérer les nématodes à galles. Maraîchage Bio. 2009; 61: 1-6.

25. Duponnois R, Ba AM, Mateille T. Effects of some rhizosphere bacteria for the biocontrol of nematodes of the genus Meloidogyne with Arthrobotrys oligospora. Fund Appl Nematol.1998; 21 (2): 157-163.

26. France. Ministère de l›Agriculture, de l›Agroalimentaire et de la Forêt. [cited 2014 sept. 04] Available from:http://agriculture.gouv.fr/

27. Fravel DR. Commercialization and implementation of biocontrol. Ann Rev Phytopathol. 2005; 43: 337-359.

28. Gaspard JT, Jaffee BA, Ferris H. Association of Verticillium chlamydosporium and Paecilomyces lilacinus with Root-knot Nematode Infested Soil. J Nematol. 1990; 22 (2): 207-213.

29. González I, Infante D, Martínez B, Arias Y, González N, Miranda I, Peteira B. Induction of chitinases and glucanases in Trichoderma spp. Strains intended for biological control. Biotecnol Appl. 2012; 29: 12-16.

30. Goswami BK, Mittal A. Management of root-knot nematode infecting tomato by Trichoderma viride and Paecilomyces lilacinus. Indian Phytopathol. 2004; 57: 235-236.

31. Goswami BK, Pandey RK, Rathour KS, Bhattacharya C, Singh L. Integrated application of some compatible biocontrol agents along with mustard oil seed cake and furadan on Meloidogyne incognita infecting tomato plants. J Zhejiang Univ Sci. B 2006; 7: 873-875.

32. Grassé PO. Traité de Zoologie, Anatomie, Systématique, Biologie. Nématodes. Vol. 4, Fasc. 2 + 3, Paris: Masson & Cie. 1965.

33. Greencorp revolucionamos la agricultura, Nemaxxion Biol. [cited 2014 sept. 04] Available from:http://greencorp.mx/productos.php?id=11&subcat=2#productos-Nemaxxion%20Biol

34. Hajieghrari B, Torabi-Giglou M, Mohammadi MR., Davari M. Biological potential of some Iranian Trichoderma isolates in control of soil borne plant pathogenic fungi. Afr J Biotechnol. 2008; 7(8): 967-972.

35. Harman GE. Changes in perceptions derived from research on Trichoderma harzianum T22. Plant Dis. 2000; 84 (4): 377-393.

36. Haseeb A, Kumar V. Management of Meloidogyne incognita-Fusarium solani disease complex in brinjal by biological control agents and organic additives. Ann Plant Protect Sci. 2006; 14: 519-521.

37. Hesseltine CW. Solid state fermentation: an overview. Int Biodeterioration. 1987; 23: 79-89.

38. Holland RJ, Williams KL, Khan A. Infection of Meloidogyne javanica by Paecilomyces lilacinus. Nematol. 1999; 1: 131-139

39. Huang XW, Zhao N.H, Zhang KQ. Extracellular enzymes serving as virulence factors in nematophagous fungi involved in infection of the host. Res Microbiol. 2004; 155 (10): 811-816.

40. Hunt DJ, Handoo ZA. Taxonomy, identification and principal species. In: Perry, RN, Moens M. Starr JL. (Eds.), Root-knot Nematodes. CABI International, Cambridge, MA (USA); 2009.

41. International Organisation for Biological Control. [cited 2014 sept. 04] Available from: http://www.iobc-global.org/about_iobc.html

42. Junaid JM, Dar NA, Bhat TA, Bhat AH, Bhat MA. Commercial Biocontrol Agents and Their Mechanism of Action in the Management of Plant Pathogens. Int J Modern Plant & Anim Sci. 2013; 1(2): 39-57.

43. Khan A, Williams KL, Nevalainen HKM. Control of plant-parasitic nematodes by Paecilomyces lilacinus and Monacrosporium lysipagum in pot trials. Biol Control. 2006; 51: 643-658.

44. Kiewnick S, Sikora RA. Biological control of the root-knot nematode Meloidogyne incognita by Paecilomyces lilacinus strain 251. Biol Control. 2006; 38: 179-187.

45. Kiewnick S, Sikora RA. Optimizing the efficacy of Paecilomyces lilacinus (strain 251) for the control of root-knot nematodes. Com Appl Biol Sc. 2004; 69: 373-380.

46. Koenning SR, Overstreet C, Noling JW, Donald PA, Becker JO, Fortnum BA. Survey of crop losses in response to phytoparasitic nematodes in the United States for 1994. J Nematol. 1999; 31: 587-618.

47. Koppert Biological Systems. Mycotal [cited 2014 sept. 04] Available from:http://www.koppert.fr/ravageurs/thrips/produits-contre-les-thrips/detail/mycotal-4

48. Koppert Biological Systems. Trianum(r) [cited 2014 sept. 04] Available from: http://www.koppert.fr/trianum/

49. Küçük C, Kivanç M. Effect of formulation on the viabilıty of biocontrol agent, Trichoderma harzianum conidia. Afr J Biotechnol. 2005; 4 (5): 483-486.

50. Kumar D, Singh KP. Assessment of predacity and efficacy of Arthrobotrys dactyloides for Biological Control of root knot disease of tomato. J Phytopathol. 2006; 154: 1-5.

51. Laverlam. Biostat(r) [cited 2014 sept. 04] Available from: http://www.laverlamintl.com/?page_id=242

52. Laverlam. BotaniGard(r) ES [cited 2014 sept. 04] Available from: http://www.laverlamintl.com/?page_id=207

53. Laverlam. Mycotrol O(r) [cited 2014 sept. 04] Available from: http://www.laverlamintl.com/?page_id=91

54. Lewis JA, Papavizas GC. Characterization of alginate pellets formulated with Trichoderma and Gliocladium and their effect on the proliferation of the fungi in soil. Plant Pathol. 1985; 34: 571-577.

55. Lewis JA, Papavizas GC. Production of chlamydospore and conidia by Trichoderma spp. liquid and solid growth media. Soil Biol Biochem. 1983; 15: 351-357.

56. Liu T, Wang L, Duan YX, Wang X. Nematicidal activity of culture filtrate of Beauveria bassiana against Meloidogyne hapla. World J Microbiol Biotechnol. 2008; 24: 113-118.

57. Lumsden RD, Lewis JA, Locke JC. Managing soilborne plant pathogens with fungal antagonists. In : RD Lumdsen, JL Vaughn, eds. Pest Management : Biologically Based Technologies. Beltsville symp XVIII. Washington, D.C: Am Chem Soc. 1993.

58. Mahmoud MF. Pathogenicity of Three Commercial Products of Entomopathogenic Fungi, Beauveria bassiana, Metarhizum anisopilae and Lecanicillium lecanii against Adults of Olive Fly, Bactrocera oleae (Gmelin) (Diptera:Tephritidae) in the Laboratory. Plant Protect Sci. 2009; 45 (3): 98-102.

59. Masadeh B, Von Alten H., Grunewald-Stoecker G., Sikora R.A. Biocontrol of root-knot nematodes using the arbuscular mycorrhizal fungus Glomus intraradices and the antagonist Trichoderma viride in two tomato cultivars differing in their

suitability as hosts for the nematodes. J Plant Dis Protect. 2004; 111, 322-333.

60. Mateille T. Pasteuria penetrans : un nouvel «outil biologique» de lutte contre les nématodes phytoparasites : perspectives d›application en cultures maraîchères. Congad Infos. 1993; (15): 1-2.

61. Mendoza AR, Sikora RA, Kiewnick S. Influence of Paecilomyces lilacinus 251 on the biological control of the burrowing nematode Radopholus similis in banana. Nematropica 2007; 37 (2): 203-215.

62. Meyer SLF. Efficacy of the fungus Verticillium lecanii for suppressing root-knot nematode egg numbers on cantaloupe roots. Hort Technology. 1999; 9: 443-447.

63. Mitchell DA, Berovic M, Krieger N. Overview of solid state bioprocessing. Biotechnol. Ann Rev. 2002; 8: 183-225.

64. Monte E, Llobell A. Trichoderma in organic agriculture. Proceedings V World Avocado Congress (Actas V Congreso Mundial del Aguacate) 2001.

65. Mukhtar T, Arshad MA, Zameer Kayani M. Biocontrol potential of Pasteuria penetrans, Pochonia chlamydosporia, Paecilomyces lilacinus and Trichoderma harzianum against Meloidogyne incognita in okra. Phytopathol Medit. 2013; 52 (1): 66-76.

66. Netscher G. Les nématodes parasites des cultures maraîchères au Sénégal. Cahiers de l›ORSTOM. 1970; 11: 209-229.

67. Novozymes. Met 52, l›insecticide biologique pour le contrôle de l›otiorhynque de la vigne [cited 2014 sept. 04] Available from:http://bioag.novozymes.com/en/products/europe/biocontrol/Documents/Europe_Met52_fact_sheet_French.pdf

68. Organic Materials Review Institute. [cited 2014 sept. 04] Available from: http://www.omri.org/

69. Papavizas GC, Fravel DR, Lewis JA. Proliferation of Taleromyces flavus in soil and survival in alginate pellets. Phytopathol.1987; 77: 131-136.

70. Prophyta. BioAct/Melocon for the control of plant parasitic nematodes [cited 2014 sept. 04] Available from:http://195.98.196.121/en/technology/nematode-diseases/

71. Quimby PC, Birdsall JL, Caesar AJ, Zidack NK, Boyette CD, Connick WJ, Grey WE. Granulated formulation and method for stabilizing biocontrol agents. U.S. Patent Application No. 8,695,249. 1996.

72. Roussos S, Raimbault M, Prebois JP, Lonsane BK. Zymotis, a large scale solid state fermenter. Appl Biochem Biotech. 1993; 42: 36-52.

73. Samson RA, Hoekstra ES, Frisvad JC, Filtenborg O. Introduction to food borne fungi. Fifth edition CBS, Netherlands. 1996.

74. Samuels GJ. Trichoderma: a review of biology and systematics of the genus. Mycol Res. 1996 ; 100: 923-935.

75. Santé Canada. [cited 2014 sept. 04] Available from: http://www.hc-sc.gc.ca/cps-spc/pest/index-fra.php

76. Sarhy-Bagnon V, Lozano P, Saucedo-Castaneda G, Roussos S. Production of 6-pentyl-alpha-pyrone by Trichoderma harzianum in liquid and solid state cultures. Process Biochem. 2000; 36: 103-109.

77. Sawadogo A, Diop MT, Thio B, Konate YA, Mateille T. Incidence de quelques facteurs agronomiques sur les populations de Meloidogyne spp. et leurs principaux organismes parasites en culture maraîchère sahélienne. Nematol. 2000; 2 (8): 895-906.

78. Selvakumar R, Srivastava KD. Management of spot blotch of wheat using bio control agent: National conference on Swadeshi Vighan, held at New Delhi, December 26-28, 2000.

79. Serp D, Cantana E, Heinzen C, Stockear VU, Marison IW. Characterization of an encapsulation device for the production of monodisperse alginate beads for cell immobilization. Biotechnol Bioengr. 2000; 70: 41-53.

80. Sharon E, Bar-Eyal M, Chet I, Herrera-Estrella A, Kleifeld O, Spiegel Y. Biological Control of the Root-Knot Nematode Meloidogyne javanica by Trichoderma harzianum. Biol Control. 2001; 91 (7): 687-693.

81. Siddiqui ZA, Mahmood I. Biological control of plant parasitic nematodes by fungi: a review. Bioresource Technol. 1996; 58: 229-239.

82. Stirling GR, Licastro KA, West LM, Smith LJ. Development of Commercially Acceptable Formulations of the Nematophagous

Fungus Verticillium chlamydosporium. Biol Contr. 1998; 11: 217-223.

83. Stirling GR, Smith LJ. Field tests of formulated products containing either Verticillium chlamydosporium or Arthrobotrys dactyloides for biological control of root-knot nematodes. Biol Contr. 1998; 11: 229-237.

84. Stirling GR. ‹›Biological Control of Plant Parasitic Nematodes.›› CAB International, Wallingford, UK. 1991.

85. Thakur NSA, Devi G. Management of Meloidogyne incognita attacking okra by nematophagous fungi, Arthrobotrys oligospora and Paecilomyces lilacinus. Agric Sci Dig. 2007; 27: 50-52.

86. United States Environmental Protection Agency. [cited 2014 sept. 04] Available from:http://www2.epa.gov/aboutepa/our-mission-and-what-we-do

87. Van Damme V, Hoedekie A, Viaene N. Long-term efficacy of Pochonia chlamydosporia for management of Meloidogyne javanica in glasshouse crops. Nematology. 2005; 7: 741-745.

88. Van der Putten WH, Cook R, Costa S, Davies KG, Fargette M, Freitas H, et al. Nematode interactions in nature: models for sustainable control of nematode pests of crop plants? Adv Agron. 2006; 89: 227-260.

89. Verdejo-Lucas S, Sorribas FJ, Ornat C, Galeano M. Evaluating Pochonia chlamydosporia in a double-cropping system of lettuce and tomato in plastic houses infested with Meloidogyne javanica. Plant Pathol. 2003; 52: 521-528.

90. Viaene NM, Abawi GS. Management of Meloidogyne hapla on lettuce in organic soil with Sudan grass as a cover crop. Plant Dis. 1998; 82: 945-952.

91. Walker HL, Connick WJ. Sodium alginate for production and formulation of mycoherbicides. Weed Sci. 1983; 31: 333-338.

92. Widden P, Abitbol JJ. Seasonality of Trichoderma species in a spruce forest soil. Mycologia. 1980; 72 (4): 775-784.

93. Zaidi WN, Singh US. Development of improved technology for mass multiplication and delivery of fungal (Trichoderma) and bacterial (Pseudomonas) Bio-Control agents. J Mycol Pl Path. 2004; 34 (3): 732-741.

Chapter 3

Use of the Arbuscular Mycorrhizal Fungus Glomus Intraradices as Biological Control Agent of the Nematode Nacobbus Aberrans Parasitizing Tomato

Nicolás Marro[1], Paola Lax[2], Marta Cabello[3], Marcelo Edmundo Doucet[2], and Alejandra Gabriela Becerra[1]

[1]Laboratorio de Micología; Instituto Multidisciplinario de Biología Vegetal; CONICET- Universidad Nacional de Córdoba; Córdoba - Argentina

[2]Instituto de Diversidad y Ecología Animal (CONICET-UNC) y Centro de Zoología Aplicada, Facultad de Ciencias Exactas, Físicas y Naturales, Universidad Nacional de Córdoba, Argentina

[3]Instituto Spegazzini, Facultad de Ciencias Naturales y Museo, Buenos Aires, Argentina

ABSTRACT

The plant-parasitic nematode Nacobbus aberrans is an endoparasite that induces gall formation in the roots and causes severe losses to diverse crops. Some populations of this nematode show preference for certain hosts, revealing the existence of "races/groups" with different behaviour and making nematode management difficult. A possible biological control alternative to reduce the damage caused by this species may be the use of arbuscular mycorrhizal fungi (AMF). In the present work, the effect of Glomus intraradices on tomato plants inoculated with the nematode at transplanting and three weeks later was tested. At 60 days, the following parameters were estimated: percentage of AMF colonization, root and aerial dry weight, number of galls and egg masses, and reproduction factor (RF=final population/ initial population) of N. aberrans. AMF colonization was higher in the presence of the nematode. The use of AMF favoured tomato biomass and reduced the number of galls and RF on the plants inoculated with the nematode at transplanting.

INTRODUCTION

Some species of plant-parasitic nematodes can cause severe damage to numerous crops. They attack mainly the root system. As a consequence, the plant loses its capacity to take up water and nutrients, with likely reductions in the growth and yield (Talavera et al. 2001). *Nacobbus aberrans* (Thorne) Thorne and Allen is a sendentary endoparasite that causes histological alterations to the root tissues, inducing gall formation. Because the galls are very similar to those produced by *Meloidogyne* spp. (root-knot nematode), *N. aberrans* is also known as the "false root-knot nematode". The nematode is native to the Americas (Sher 1970), and is present in Argentina, Bolivia, Chile, Ecuador, Mexico, Peru, and the United States. It is a species of quarantine importance (Karssen and den Nijs 2001) that has a wide host range. Parasitized plants can show poor development, chlorotic leaves, wilting and small-sized fruits (Manzanilla-López et al. 2002). *N. aberrans* was first detected in north-western Argentina in 1977 parasitizing several crops and some weeds (Costilla et al. 1977). At present, it has a wide distribution (Doucet and Lax 2005). In Córdoba province, it causes severe damage to tomato

(*Solanum lycopersicum* L.) and pepper (*Capsicum annuum* L.), both crops being grown in greenhouses (Lax et al. 2011).

Arbuscular mycorrhizal fungi (AMF) establish a symbiotic relationship with most plant species (Wang and Qiu 2006). These fungi increase host development and tolerance against several soil pathogens, such as plant-parasitic nematodes. AMF facilitate uptake of the nutrients, mainly of phosphorus (Borowicz 2001), increasing plant vigor (Azcón-Aguilar and Barea 1996). In addition, AMF affects the nematodes by altering root exudates (Graham and Menge 1982) or by producing phytoalexins in the root, such as glyceollin, which could have nematicidal effects (Morandi 1996). Although possible mechanisms have been proposed to explain the reduction of harmful nematodes in the presence of AMF, they remain a matter of speculation (Elsen et al. 2001; Elsen et al. 2003).

The interaction between the AMF and *Meloidogyne* spp. in the plants of economic importance has been widely documented, with records of reduction in reproduction (Jaizme-Vega et al. 1997; Castillo et al. 2006), number of second-stage juveniles (J2) in the soil (Shreenivasa et al. 2007; Gómez et al. 2008) and number of egg masses (Da Silva Sousa et al. 2010). The above works also reported a lower number of galls induced by these nematodes.

Studies on *N. aberrans*-AMF interaction are scarce (Gardezi et al. 1995; Lax et al. 2011). A reduced number of galls was detected in tomato plants colonized by *Glomus* spp. (Gardezi et al. 1995). Tomato plants colonized by *G. intraradices* Schenck and Smith and further inoculated with an Argentine population of the nematode showed a reduction in the multiplication of the parasite (Lax et al. 2011). *N. aberrans* populations can behave in a different way on a single plant or a range of hosts; hence, the existence of "races/groups" within the species is considered (Manzanilla-López et al. 2002). Such physiological variability hinders the agricultural damage produced by the nematode. Therefore, studies with new populations from different geographical regions are necessary. The aim of the present work was to evaluate the effect of *G. intraradices* on a new and different Argentine population of *N. aberrans* and on tomato plant biomass.

MATERIALS AND METHODS

A population of *N. aberrans* from the locality of Río Cuarto (Río Cuarto department, Córdoba province, Argentina) was used for this study. Nematodes were multiplied on tomato cultivar Platense under greenhouse conditions. Egg masses of the nematode were extracted from the infected plants and placed in Petri dishes containing sterile water. They were maintained at room temperature to allow hatching; then, mobile J2s were recovered.

Spores and roots colonized by *G. intraradices* (origin: La Plata, Herbario Spegazzini (LPS), Tierra del Fuego 28) were used for the study. The fungus was propagated on leek (*Allium porrum* L.) in the pots containing a mixture of sterile perlite and vermiculite (1:1) for six months. After this period, roots (0.25 g) were chopped into 1-cm long segments to determine the mycorrhizal colonization (Phillips and Hayman 1970) and quantify the AMF (Giovannetti and Mosse 1980). To obtain the spores, *G. intraradices* was propagated on alfalfa (*Medicago sativa* L.) plants for six months in the pots containing the same substrate as used for leek and fed with a nutrient solution three times a week (Cabello 1997). After that, a sample (5.0 g) of the perlite-vermiculite mixture was extracted to quantify the number of spores using the wet sieving and decanting technique (Gerdemann and Nicolson 1963).

The seeds of tomato Platense were surface-sterilized with 10% NaClO for five minutes and placed to germinate in the trays containing a mixture of sterile soil and sand (3:1). The physical and chemical characteristics of the soil were as follows (%): silt 23, clay 20, sand 57, organic matter 13.2, N 0.5, and P 21 mg/Kg; pH 5.2 One-month-old seedlings were transplanted to the containers (20 cm long x 4 cm wide, with a capacity of 190 g of soil). The experimental design consisted of six treatments with six replicates each: 1) non-inoculated plants (Control), 2) plants inoculated with *G. intraradices* at transplanting (AMF T0), 3) plants inoculated with *N. aberrans* at T0 (NEM T0), 4) plants inoculated with *N. aberrans* and *G. intraradices* at T0 (AMF T0 + NEM T0), 5) plants inoculated with *N. aberrans* three weeks after transplanting (NEM T3), 6) plants inoculated with *G. intraradices* at T0 and with *N. aberrans* three weeks after transplanting (AMF T0 + NEM T3).

At transplanting, the container was filled to two-third of its capacity with the substrate. In the treatments with*N. aberrans*, the initial population (Pi) of the nematode (100 J2/1.5 mL of sterile water) was deposited on the roots. In the plants treated with AMF, a homogeneous layer of the fungal inoculum (0.25 g of leek roots with 62.5% colonization and 5.0 g of the perlite-vermiculite mixture containing 1800 spores) was deposited; tomato seedlings were placed on the substrate and then roots were covered with the substrate. Plants were grown in a greenhouse (24/18°C day/night, 16/8 h day/night photoperiod). After 60 days, the plants were removed and biomass (aerial and root dry weight) was measured. The number of galls induced by the nematode was counted. Egg masses were extracted, counted and placed in a 1% NaClO solution for 4 min to count the eggs (Hussey and Barker 1973). The soil of each plant was processed using the centrifugal-flotation technique (Jenkins 1964) to obtain vermiform stages. Final population (Pf) was estimated for each replicate by adding the number of eggs and nematodes extracted from the soil. Pf value was used to calculate the Reproduction Factor (RF=Pf/Pi). In addition, for all the treatments, roots were stained and total mycorrhizal colonization was quantified, following the previously mentioned techniques.

Normality and homogeneity assumptions were confirmed for all the variables before statistical analyses. An ANOVA was performed to detect the differences among the treatments ($p<0.05$). The DGC multiple comparison test (test of Di Rienzo, Guzmán and Casanoves) was applied (Infostat 2001).

RESULTS

Both plant growth parameters showed significant differences among the treatments (Table 1). Aerial biomass was higher in AMF T0 and NEM T0 than in the control. Root biomass increased in AMF T0 compared to non-inoculated plants and NEM T0. Plants of the treatment AMF T0 + NEM T0 showed a greater aerial and root biomass than the control and NEM T0. No significant differences were observed between the two treatments inoculated with *N. aberrans* at T3 with respect to the control.

Table 1: Dry weight of tomato (cultivar Platense) inoculated with Glomus intraradices (AMF) and Nacobbus aberrans (NEM)

Treatment	Aerial part (g)	Root (g)
Control	0.43 ± 0.13c1	0.24 ± 0.09b
AMF T0	0.89 ± 0.12a	0.45 ± 0.15a
NEM T0	0.63 ± 0.2b	0.19 ± 0.03b
AMF T0 + NEM T0	0.77 ± 0.13a	0.46 ± 0.14a
NEM T3	0.44 ± 0.15c	0.12 ± 0.02b
AMF T0 + NEM T3	0.34 ± 0.04c	0.14 ± 0.06b

Values are mean and standard deviation of six replicates per treatment. 1Different letters between rows indicate significant differences (p

Mycorrhizal colonization of *G. intraradices* was observed only in those treatments that were inoculated with the fungus. Root systems of tomato showed external and intracellular aseptate hyphae, terminal arbuscules, intracellular vesicles and coils (Fig. 1). Plants parasitized with the nematode showed a significant increase in mycorrhizal colonization (Table 2).

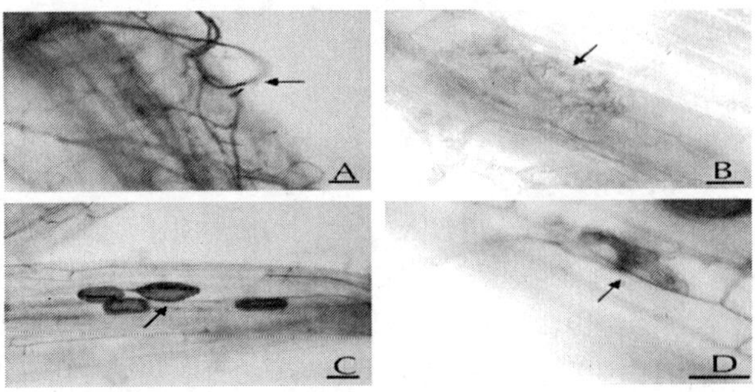

Figure 1: Mycorrhizal colonization of Glomus intraradices in tomato (cultivar Platense) roots infected with Nacobbus aberrans. A) External hyphae (arrow);

B) Arbuscule (arrow); C) Vesicles (arrow); D) Coil (arrow). Scale bar: A, C, and D: 50 μm; B: 10 μm.

Table 2: Percentage of mycorrhizal colonization of G. intraradices (AMF) in tomato (cultivar Platense) inoculated with Nacobbus aberrans (NEM)

Treatment	Colonization (%)
AMF T0	22.87 ± 6.81 b1
AMF T0 + NEM T0	35.89 ± 2.8 a
AMF T0+ NEM T3	43.96 ± 9.42 a

Values are mean and standard deviation of six replicates per treatment. 1Different letters indicate significant differences (p

All the tomato plants inoculated with *N. aberrans* J2s presented galls and egg masses generated by the nematode (Fig. 2). In NEM T0, the mean value of number of galls and RF was significantly higher than in the remaining treatments (Table 3). No significant differences were observed for number of egg masses.

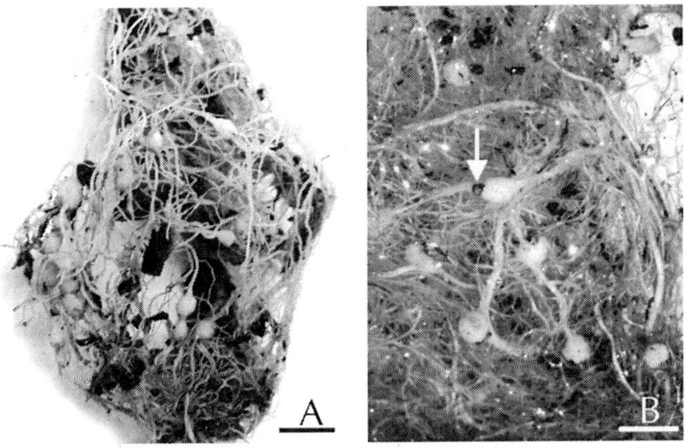

Figure 2: Tomato roots parasitized by Nacobbus aberrans. A) Galled root system (NEM T0); B) Detail of galls with egg mass (arrow). Scale bar: A and B: 1 cm.

Table 3: Effect of the application of Glomus intraradices on tomato infected by Nacobbus aberrans

Treatment	Number of galls	Number of egg masses	Reproduction factor
NEM T0	20.33 ± 6.44a1	8 ± 3.69a	36.97 ± 21.88a
AMF T0 + NEM T0	9 ± 4.77b	4.5 ± 2.35a	11.53 ± 6.23b
NEM T3	14.17 ± 6.82b	5.17 ± 2.32a	14.2 ± 9.39b
AMF T0 + NEM T3	8.83 ± 4.49b	3.83 ± 2.48a	11.44 ± 11.62b

Values are mean and standard deviation of six replicates per treatment. 1Different letters between rows indicate significant differences (p

DISCUSSION

Plant-parasitic nematodes have been traditionally managed with nematicides (Vos et al. 2012). Most of the chemical products used do not comply with current environmental regulations and soon they might be replaced with other management strategies of lower environmental impact (Harrier and Watson 2004). Improving the host tolerance through the association with mycorrhizal fungi seems to be a promising option.

In the present work, the application of *G. intraradices* at transplanting promoted the increase of aerial and root biomass of tomato, even in the presence of *N. aberrans*. An increase in biomass was observed in tomato inoculated with *G. etunicatum* Becker and Gerd. and *Gigaspora margarita* Becker and Hall, and parasitized with*M. javanica* (Treub) Chitwood (Cofcewicz et al. 2001). A similar situation was observed with *G. etunicatum* in kiwi (*Actinidia delicios* Chev.), attacked by *M. javanica* (Verdejo et al. 1990). In the present work, pre-inoculation of tomato with AMF and three weeks later with *N. aberrans* did not promote plant biomass. AMF may increase productivity by enhancing the plant resource uptake when competing with soil pathogens

(Schnitzer et al. 2011). In the plants of the last treatment (AMF T0 + NEM T3), AMF was less time in contact with *N. aberrans*. Hence, AMF would not be able to increase resource availability to host, affecting biomass increase.

The percentage of mycorrhizal colonization can increase or decrease in the presence of plant-parasitic nematodes (Roncadori 1997), depending on the arbuscular mycorrhizal species involved (Waceke et al. 2001). Here, the two treatments inoculated with nematodes and AMF (T0 and T3) showed increase in mycorrhizal level with respect to AMF at T0. *Meloidogyne incognita* (Kofoid and White) Chitwood, also showed an increase in the colonization of *Acaulospora scrobiculata* Trappe in tomato (Da Silva Sousa et al. 2010), of *G. mosseae* (Nicol. and Gerd.) Gerd. and Trappe in banana (Jaizme-Vega et al. 1997), and of *Gi. margarita* in soybean (*Glycine max* (L.) Merrill) (Carling et al. 1989). Alban et al. (2013) observed a high percentage of AMF colonization in the roots of coffee (*Coffea arabica* L.) parasitized with *M. exigua* Göldi. According to these authors, if AMF and the nematode were simultaneously present in the soil, AMF colonized the roots before nematode infestation. Thus, coffee plants can regain the energy lost by the parasitic interaction, preventing the reduction in plant growth. However, in some situations, *Glomus* spp. colonization has been lower in the presence of *Meloidogyne* spp. on different hosts (Atilano et al. 1981; Waceke et al. 2001; Elsen et al. 2002). The presence of the nematode may interfere with nutrient flow between the root and the fungus, reducing AMF efficiency (Cofcewicz et al. 2001).

Tomato was benefited by *G. intraradices* when inoculated at transplanting, through a 56% reduction in the number of galls induced by *N. aberrans*. A reduced number of galls was observed in the application of different fungus species on tomato roots attacked by *M. incognita* (Da Silva Sousa et al. 2010) and *N. aberrans* (Gardezi et al. 1995; Lax et al. 2011). Although there were no differences in the number of eggs produced by *N. aberrans* among treatments, a significant reduction (69%) in the RF of the nematode at T0 was observed. A decrease in RF of different plant-parasitic nematodes was detected in the presence of several AMF species (including *G. intraradices*) (Pinochet et al. 1998; Elsen et al. 2002; Talavera et al. 2002; Castillo et al. 2006; Lax et al. 2011; Vos et al. 2012). A reduction was observed in the final population of *Pratylenchus vulnus* Allen and Jensen in the roots of peach (*Prunus persica* L.) colonized by *G. mosseae*. This phenomenon was attributed to possible AMF-induced physiological changes in the roots, which would make the roots an unfavourable food source for the parasite (Pinochet et al. 1995).

The application of *G. intraradices* at transplanting significantly reduced the number of galls and RF of *N. aberrans*. However, in the plants mycorrhized and further inoculated with the nematode (AMF T0 + NEM T3), those parameters did not show significant differences from the respective control treatment (NEM T3). These results did not agree with the earlier studies considering *M. incognita* and *G. mosseae* on the same host (Talavera et al. 2001). The increased development of the root system three weeks after transplanting (regardless of presence or absence of mycorrhizal infection) might have affected the root penetration by juveniles, with the consequent reduction in the number of galls and final population of the nematode.

This work confirmed once again the usefulness of applying AMF at transplanting as a tool to manage *N. aberrans*populations on tomato. No tomato cultivars resistant to this nematode have been developed so far. AMF-tomato interaction might be an efficient control strategy in commercial production of this horticultural crop, especially under greenhouse conditions. In *N. aberrans*-contaminated soils, the interaction would contribute to decrease the nematode population density, therefore reducing the damage produced by the parasite and increasing the crop yield.

ACKNOWLEDGEMENTS

This work was financially supported by Secretaría de Ciencia y Técnica (Universidad Nacional de Córdoba), Ministerio de Ciencia y Tecnología (province of Córdoba), and Consejo Nacional de Investigaciones Científicas y Técnicas (CONICET).

REFERENCES

1. Alban R, Guerrero R, Toro M. Interaction between a root knot nematode (Meloidogyne exigua) and arbuscular mycorrhizae in coffee plant development (Coffea arabica). Am J Plant Sci. 2013; 4: 19-23.

2. Atilano RA, Menge JA, Van Gundy SD. Interaction between Meloidogyne arenaria and Glomus fasciculatus in grape. J Nematol. 1981; 13: 52-57.

3. Azcón-Aguilar C, Barea JM. Arbuscular mycorrhizas and biological control of soil-borne plant phatogens- an overview of the mechanisms involved. Mycorrhiza. 1996; 6: 457-464.

4. Borowicz VA. Do arbuscular mycorrhizal fungi alter plant-pathogen relations?. Ecology. 2001; 82: 3057-3068.

5. Cabello MN. Hydrocarbon pollution: its effect on native arbuscular mycorrhizal fungi (AMF). FEMS Microbiol Ecol. 1997; 22: 233-236.

6. Carling DE, Roncadori RW, Hussey RS. Interactions of vesicular-arbuscular mycorrhizal fungi, root-knot nematode, and phosphorus fertilization on soybean. Plant Dis. 1989; 73: 730-733.

7. Castillo P, Nico AI, Azcón-Aguilar C, Del Río Rincón C, Calvet C, Jimenez-Díaz RM. Protection of olive planting stocks against parasitism of root-knot nematodes by arbuscular mycorrhizal fungi. Plant Pathol. 2006; 55: 705-713.

8. Cofcewicz ET, Medeiros CAB, Carneiro RMDG, Pierobom CR. Interação dos fungos micorrízicos arbusculares Glomus etunicatum e Gigaspora margarita e o nematóide das galhas Meloidogyne javanica em tomateiro. Fitopatol Bras. 2001; 26: 65-70.

9. Costilla MA, González de Ojeda S, Hasselrot de Gomez T. Contribución al estudio del falso nematodo del nudo Nacobbus aberrans. Nematropica. 1977; 7: 7-8.

10. Da Silva Sousa C, Fermino Soares AC, Coimbra JL, Da Silva Garrido M, Da Silva Machado G. Fungos micorrízicos arbusculares no controle de Meloidogyne incognita em mudas de tomateiro. Rev Caatinga. 2010; 23: 15-20.

11. Doucet ME, Lax P. El género Nacobbus Thorne & Allen, 1944 en Argentina. 6. La especie N. aberrans (Thorne, 1935) Thorne & Allen, 1944 (Nematoda: Tylenchida) y su relación con la agricultura. An Acad Nac Agronomía Vet. 2005; 59: 5-45.

12. Elsen A, Declerck S, De Waele D. Effects of Glomus intraradices on the reproduction of the burrowing nematode (Radopholus similis) in dixenic culture. Mycorrhiza. 2001; 11: 49-59.

13. Elsen A, Declerck S, De Waele D. Effect of three arbuscular mycorrhizal fungi on root-knot nematode (Meloidogyne spp.) infection of Musa. Infomusa. 2002; 11: 21-23.
14. Elsen A, Declerck S, De Waele D. Use of root organ cultures to investigate the interaction between Glomus intraradices and Pratylenchus coffeae. Appl Environ Microbiol. 2003; 69: 4308-4311.
15. Gardezi AK, Zavaleta-Mejía E, Garcia R, Ferrera CR, Perez CA. Efecto de endomicorrizas sobre la infección de Nacobbus aberrans en jitomate (Lycopersicon esculentum). Memorias XXII Congreso Nacional de Fitopatología, Guadalajara, México. 1995; pp. 146-148. .
16. Gerdemann JW, Nicolson TH. Spores of mycorhizal Endogone species extracted from soil by wet sieving and decanting. Trans Brit Mycol Soc. 1963; 46: 235-244.
17. Giovannetti M, Mosse B. An evaluation of techniques for measuring vesicular arbuscular mycorrhizal infection in roots. New Phytol. 1980; 84: 489-500.
18. Gómez L, Rodríguez MG, De la Noval B, Miranda I, Hernández MA. Interacción entre el Ecomic(r) y una población cubana de Meloidogyne incognita en tomate. Rev Protección Veg. 2008; 23: 90-98.
19. Graham JH, Menge JA. Influence of leaf phosphorus on take-all disease of wheat. Phytopatology. 1982; 72: 95-98.
20. Harrier LA, Watson CA. The potential role of arbuscular mycorrhizal (AM) fungi in the bioprotection of plants against soil-borne pathogens in organic and/or other sustainable farming systems. Pest Manag Sci. 2004; 60: 149-157.
21. Hussey RS, Barker KR. A comparation of methods of collecting inocula of Meloidogyne spp., including a new technique. Plant Dis Rep. 1973; 57: 1025-1028.
22. Infostat. Programa de análisis estadísticos. Cátedra de estadística. Facultad de Ciencias Agropecuarias, Universidad Nacional de Córdoba. 2001.
23. Jaizme-Vega MC, Tenoury P, Pinochet J, Jaumot M. Interactions between the root-knot nematode Meloidogyne incognita and Glomus mosseae in banana. Plant Soil. 1997; 196: 27-35.

24. Jenkins WR. A rapid centrifugal-flotation technique for separating nematodes from soil. Plant Dis Rep. 1964; 48: 692.
25. Karssen G, den Nijs LJMF. Diagnostic protocols for regulated pest: Nacobbus aberrans. EPPO Bulletin. 2001; 3: 71-77.
26. Lax P, Becerra AG, Soteras F, Cabello M, Doucet ME. Effect of the arbuscular mycorrhizal fungus Glomus intraradices on the false root-knot nematode Nacobbus aberrans in tomato plants. Biol Fertil Soils. 2011; 47: 591-597.
27. Manzanilla-López RH, Costilla MA, Doucet M, Franco J, Inserra RN, Lehman PS, et al. The genus Nacobbus Thorne & Allen, 1944 (Nematoda: Pratylenchidae): systematics, distribution, biology and management. Nematropica. 2002; 32: 149-227.
28. Morandi D. Ocurrence of phytoalexins and phenolic compounds in endomycorrhizal interactions, and their potencial role in biological control. Plant Soil. 1996; 185: 241-251.
29. Phillips JM, Hayman DS. Improved procedures for clearing roots and staining parasitic and vesicular-arbuscular mycorrhizal fungi for rapid assessment of infection. Trans Brit Mycol Soc. 1970; 55: 158-161.
30. Pinochet J, Calvet C, Camprubi A, Fernandez C. Growth and nutritional response of Nemared peach rootstock infected with Pratylenchus vulnus and the mycorrhizal fungus Glomus mosseae. Fundam Appl Nematol. 1995; 18: 205-210.
31. Pinochet J, Camprubí A, Calvet C, Fernández C. Inducing tolerance to the root-lesion nematode Pratylenchus vulnus by early mycorrhizal inoculation of micropropagated Myrobalan 29 C plum rootstock. J Amer Soc Hort Sci. 1998; 123: 342-347.
32. Roncadori RW. Interactions between arbuscular mycorrhizas and plant parasitic nematodes in agro-ecosystems. In: Gange AC, Brown VK. Multitrophic interactions in terrestrial systems: 36th Symposium of the British Ecological Society. Blackwell, Oxford, England, 1997; pp. 101-113.
33. Schnitzer SA, Klironomos JN, HilleRisLambers J, Kinkel LL, Reich PB, Xiao K, et al. Soil microbes drive the classic plant diversity-productivity pattern. Ecology. 2011; 92: 296-303.
34. Sher SA. Revision of the genus Nacobbus Thorne and Allen, 1944 (Nematoda: Tylenchoidea). J Nematol. 1970; 2: 228-235.

35. Shreenivasa KR, Krishnappa K, Ravichandra NG. Interaction effects of arbuscular mycorrhizal fungus Glomus fasciculatum and root-knot nematode, Meloidogyne incognita on growth and phosphorous uptake of tomato. Karnataka J Agric Sci. 2007; 20: 57-61.

36. Talavera M, Itou K, Mizukubo T. Reduction of nematode damage by root colonization with arbuscular mycorrhiza (Glomus spp.) in tomato-Meloidogyne incognita (Tylenchida: Meloidogynidae) and carrot-Pratylenchus penetrans (Tylenchida: Pratylenchidae) pathosystems. Appl Entomol Zool. 2001; 36: 387-392.

37. Talavera M, Itou K, Mizukubo T. Combined application of Glomus sp. and Pasteuria penetrans for reducing Meloidogyne incognita (Tylenchida: Meloidogynidae) populations and improving tomato growth. Appl Entomol Zool. 2002; 37: 61-67.

38. Verdejo S, Calvet C, Pinochet J. Efecto de la micorrización en kiwi infestado por los nematodos Meloidogyne hapla y M. javanica. Bol San Veg Plagas. 1990; 16: 629-624.

39. Vos C, Geerinckx K, Mkandawire R, Panis B, De Waele D, Elsen A. Arbuscular mycorrhizal fungi affect both penetration and further life stage development of root-knot nematodes in tomato. Mycorrhiza. 2012; 22: 157-163.

40. Waceke JW, Waudo SW, Sikora R. Response of Meloidogyne hapla to mycorrhiza fungi inoculation on pyrethrum. Afr J Sci Technol. 2001; 2: 63-70.

41. Wang B, Qiu YL. Phylogenetic distribution and evolution of mycorrhizas in land plants. Mycorrhiza. 2006; 16: 299-363.

Chapter 4

Ecological Complexity and the Success of Fungal Biological Control Agents

Guy R. Knudsen[1], and Louise-Marie C. Dandurand[2]

[1]Soil & Land Resources Division, University of Idaho, Moscow, ID 83844, USA

[2]Department of Plant, Soil, & Entomological Sciences, University of Idaho, Moscow, ID

ABSTRACT

Fungal biological control agents against plant pathogens, especially those in soil, operate within physically, biologically, and spatially complex systems by means of a variety of trophic and nontrophic interspecific interactions. However, the biocontrol agents themselves are also subject to the same types of interactions, which may reduce or in some cases enhance their efficacy against target plant pathogens. Characterization of these ecologically complex systems is challenging, but a number of tools are available to help unravel this complexity. Several of these tools are described here, including the use of molecular biology to generate biocontrol agents with useful marker genes and then to quantify these agents in natural systems, epifluorescence and confocal laser scanning microscopy to observe their presence and activity in situ, and spatial statistics and computer simulation modeling to evaluate and predict these activities in heterogeneous soil habitats.

INTRODUCTION

Reduction of chemical pesticide usage, including chemicals for control of soilborne plant pathogens, is widely recognized as a desirable goal for agriculture and forestry. Many traditional chemical means of disease control (e.g., methyl bromide and certain fungicides) are being or have been phased out due to economic considerations and/or mandate. Application of fungi and bacteria as microbial antagonists of plant pathogens offers prospects of environmentally benign pest control. A wide range of fungi have shown potential as agents for biological control of soilborne phytopathogens, and this review will focus on these fungal biocontrol agents.

Unlike gnotobiotic systems, soil is characterized by a multitude of both trophic (food webs) and nontrophic (e.g., mutualism, commensalism, neutralism, amensalism, antagonism, and competition) interspecific relationships [1]. The distinction between trophic and nontrophic relationships is somewhat arbitrary, since it focuses only on "what an organism eats and what eats it." Parasitism typically is categorized as a trophic interaction, since a parasite derives nutrition from its host, whereas mutualism is generally considered a nontrophic interaction. Nonetheless, in an alternate taxonomy of interspecific relationships, mutualism and parasitism are often considered to be part of a continuum [2, 3].

Biocontrol microbes that are applied to seeds or soil prior to planting may colonize the spermosphere and/or rhizosphere of seedlings and thus may be present at or near infection courts of soilborne pathogens. There, biocontrol agents may participate in a variety of trophic and nontrophic interaction mechanisms including production of antifungal compounds, hyperparasitism of pathogens, stimulation of host plant defenses, or competitive colonization of spermosphere and rhizosphere substrates. However, soils are both densely populated and highly competitive habitats, and poor competitive ability of introduced fungal agents in soil communities, leading to unsuccessful establishment and activity, has often been observed [4].

Understanding effects of biodiversity on soil ecosystem functions requires integration both within trophic levels (i.e., horizontal diversity, the number of species within a trophic level) and across trophic levels (vertical diversity, the number of trophic levels) [5, 6]. The significance of both horizontal diversity and vertical diversity, from an agronomic perspective, may vary with the system being considered. A number of soilborne fungi (e.g., some members of the genera Fusarium, Rhizoctonia, Pythium, Sclerotinia, and others) are pathogenic on roots of both crop plants and weeds. Thus, they may be viewed either as harmful pathogens or as beneficial biocontrol agents, depending on the agronomic significance of the plant host. For example, Sclerotinia sclerotiorum and S. minor are important crop pathogens and also have been reported as weed biocontrol agents: S. sclerotiorum against knapweed [7] and S. minoragainst dandelion [8].

When analysis is limited only to host and parasite, the system may be horizontally diverse (several potential pathogens) but not vertically diverse (only two trophic levels: host and parasite). Studies of Sclerotinia as a weed biocontrol agent might typically be less vertically diverse (two trophic levels: host and parasite) compared to attempts at biological control of phytopathogenic Sclerotinia with a fungal hyperparasite such as Trichoderma (three trophic levels: crop plant, parasite, and biocontrol hyperparasite). Screening a large number of potential antagonists to a plant pathogen is an exercise that focuses exclusively on horizontal diversity. However, failure to account for nontrophic interactions and for trophic vertical diversity is likely one reason why promising potential biocontrol agents often fail to live up to expectations in the field. As one example, fluorescent pseudomonads and other bacteria have been shown to inhibit and lyse germlings of Trichoderma spp. [9]. We observed that a soil pseudomonad, Pseudomonas fluorescens isolate 2–79, significantly inhibited radial growth and hyphal density of the biocontrol fungus T. harzianum in soil [10]. In another study, we isolated a fungivorous nematode (Aphelenchoides sp.) from field soil and found that the nematode significantly reduced hyphal growth and total population of T. harzianum under soil environmental conditions favorable to nematode activity [11]. Because hyphal growth and proliferation are an attribute of the biocontrol fungus that appears to correlate with its success as a biocontrol agent, consideration of such nontrophic and vertically diverse trophic interactions may provide insight into the "failure parameters" that often plague biocontrol efforts.

Although numerous studies have suggested that, for effective biocontrol activity, successful hyphal growth and proliferation of the agent in soil or the rhizosphere are critical, the presence of an indigenous fungal microbiota makes this difficult to evaluate. As noted by Martin and English [12], the structural, physical, and biological complexity of the soil environment in which pathogens interact with plant roots constrain disease control options, including biological control. As seeds germinate and roots subsequently elongate, the spatial and temporal availability of infection courts are constantly changing. Colonization dynamics of biocontrol agents on seeds and roots, in the presence of an indigenous rhizosphere microbial community, will determine how well these potential infection courts are protected. One of the long-term goals of our research has been to understand and quantify some of the biotic and abiotic parameters that determine the establishment and activity of fungal agents introduced into natural habitats.

Certain organisms have received considerable research attention as potential biocontrol agents; one of the most well studied of these is the fungal genus Trichoderma. Trichoderma spp. have received considerable attention as potential biological control agents against a wide range of soilborne plant pathogenic fungi [13, 14]. Members of the genus produce a number of fungal cell wall-degrading enzymes including chitinases, 1,3-glucanases, proteases, mannanases, and other hydrolases [15, 16]. Trichoderma spp. are highly ecologically successful fungi and have been used increasingly in commercial agriculture, especially in the greenhouse industry where a significant proportion of soilborne disease control products are based on T. harzianum [15].

Isolates of Trichoderma harzianum have been reported as antagonists of several soilborne plant pathogens. An Idaho isolate of T. harzianum, ThzID1, controlled Fusarium wilt of Douglas-fir seedlings [17] and colonized sclerotia of Sclerotinia sclerotiorum under laboratory and field conditions [10, 18, 19]. Experimental biocontrol of the important pea pathogens Pythium ultimum and Aphanomyces euteiches f. sp. pisi has been obtained with agents including several bacterial species and also members of the genus Trichoderma [9, 18, 20–24]. Seed treatment of pea and other crops with T. harzianum controlled Pythium seed rot and/or preemergence damping-off [25–27].

Trichoderma spp. also have shown considerable potential for bioremediation activity in soil, especially for hydrocarbon pollutants, and many of the same ecological constraints on efficacy may apply in bioremediation systems. Trichoderma spp. and especially T. harzianum can degrade a number of soil pollutants; for example, Katayama and Matsumura [28] reported on the ability of T. harzianum to degrade organochlorine pesticides, particularly endosulfan. Ermisch and Rehm [29] described degradation of polycyclic aromatic hydrocarbons, including anthracene, by T. harzianum in soil. Because of the wealth of background information onTrichoderma spp. as both biocontrol and bioremediation agents, the genus is an excellent model for the investigation of factors influencing survival and proliferation of recombinant fungi introduced into the soil.

COMPLEX SOIL MICROBIAL INTERACTIONS: NEW TOOLS AND OLD CHALLENGES

Molecular and Microscopic Tools

Several factors constrain the observation, quantification, and evaluation of biocontrol fungi in soil. Although various methods have been used to study the occurrence and distribution of Trichoderma in natural soils [30–32], relatively few methods have allowed quantitative evaluation of population dynamics and proliferation. For example, in efforts to quantify the influence of temperature, soil matric potential, nutrient source, and antagonistic bacteria, on the hyphal growth and biocontrol efficacy of pelletized Trichoderma harzianum, it was not possible to differentiate the hyphal growth of this fungal agent from indigenous Trichodermastrains in natural soils [33, 34]. The use of dilution plating for numerical estimation of fungal population does not differentiate among the different propagules (hyphal fragments, conidia, and chlamydospores) that may generate colonies when plated on agar and thus is an unreliable estimate of fungal biomass and active physiological status [35]. The use of mutant strains resistant to specific fungicides may partially overcome problems related to nonspecific recovery [36], but this method does not allow for in situ monitoring of growth dynamics and survival structures of introduced Trichoderma strains or differentiation of introducedTrichoderma strains from indigenous strains.

More recently, genetic engineering of biocontrol agents with reporter or marker genes has provided useful tools for detection and monitoring of introduced biocontrol agents in natural environments [31, 37]. The selectable hygromycin B (hygB) phosphotransferase gene, encoding for resistance to this antibiotic, has been used to monitor survival of biocontrol agents in the rhizosphere or on the phylloplane [37, 38]. The -glucuronidase (GUS) reporter gene also is a promising tool for ecological studies of biocontrol agents [31]. Advantages of this reporter gene system include the low background activity of GUS in fungi and plants, the relative ease and sensitivity of detection [38], and the apparent lack of influence of GUS expression on biocontrol efficacy [39]. However, some GUS activity may be present in unsterile systems or natural soils. For example, Aspergillus niger has some indigenous GUS activity [39]. Therefore, for study of growth patterns of an introduced fungus in natural ecosystems, this reporter gene system may be less useful. The green fluorescent protein gene (GFP) of the jellyfish Aequorea victoria also has been developed as a reporter for gene expression [40]. Expression of cloned GFP has been reported in several organisms [41, 42]. GFP was shown to be a useful tool for studying host-fungal pathogen interactions in vivo [43] and has been used to assess colonization and dispersion of Aureobasidium pullulans in the phyllosphere [44]. GFP requires only UV or blue light and oxygen to induce green fluorescence. An exogenous substrate, which GUS requires, is not needed for the detection system to function, thus avoiding problems related to cell permeability and substrate uptake [42].

In our lab, we were able to use a stable transformant of T. harzianum expressing both GFP and GUS phenotypes [45] to evaluate aspects of the ecology of this organism in soil. The fungus was cotransformed with GFP, GUS, and hygromycin B (hygB) resistance, using PEG-mediated transformation. One cotransformant (strain ThzID1-M3) has been phenotypically stable with respect to GFP expression for several years, despite repeated subculturing without selection pressure. Morphology of ThzID1-M3 is similar to that of the wild-type strain. In nonsterile soil, growth, sporulation, and the ability of the cotransformant to colonize sclerotia ofSclerotinia sclerotiorum were compared with the wild-type strain. Wild-type and transformant strains both increased colonization of sclerotia above background levels of colonization by indigenous Trichoderma spp. in untreated controls. There were no significant differences in colonization levels between wild-type and cotransformant strains; however, presence of the GFP and GUS marker genes in the cotransformant strain allowed differentiation of introducedTrichoderma from indigenous strains. Formation of greenfluorescing conidiophores and conidia was observed within the first three days of incubation in soil, followed by formation of terminal and intercalary chlamydospores and subsequent disintegration of older hyphal segments. Addition of X-Gluc substrate to recovered glass slides confirmed the activity of GUS as well as GFP in the cotransformant in soil; the green color of cotransformant hyphae was clearly visible under UV epifluorescence, while indigenous fungi in the same samples were barely visible microscopically. In one study [11], interactions between ThzID1-M3 and the fungivorous nematode Aphelenchoides sp. were investigated in untreated field soil using epifluorescence microscopy; the GFP phenotype allowed for effective identification and quantification of hyphae of ThzID1-M3 in nonsterile soil. We have also used this method to evaluate effects of indigenous soil microflora on biomass and biocontrol efficacy of ThzID1-M3 in soil [46]. Combining epifluorescence with computer image analysis allowed quantification of mycelial biomass, over time, of the GFP-transformed ThzID1-M3 following its introduction into nonsterile soil [47].

Real-time PCR amplification of Trichoderma spp. gene sequences allowed quantification of sclerotia of S. sclerotiorum by Trichoderma spp. in nonsterile soil [48]. The fact that T. harzianum strain ThzID1-M3 is phenotypically different (in that it expresses GFP) from other soil fungi (including other Trichoderma strains) and that also the corresponding novel gene is well characterized provided another avenue of ecological investigation, since the GFP gene (which is of jellyfish origin) provided a unique sequence to design a polymerase chain reaction (PCR) primer/probe set to exclusively amplify that sequence from ThzID1-M3 [49]. We also compared quantitative real-time PCR with confocal microscopy to evaluate colonization of sclerotia by ThzID1-M3 and concluded that microscopy combined with image analysis provides better information on the spatial and temporal dynamics of sclerotial colonization, while real-time PCR can provide a more precise assessment of the total extent of colonization over time, including biomass estimates for the biocontrol fungus, and can more easily be used to sample entire sclerotia [50]. Real-time PCR primers were designed for the pea pathogenFusarium solani f. sp. pisi, allowing for quantitative assessment of competitive interactions between the biocontrol fungus (T. harzianum ThzID1-M3) and the pathogen F. solani in the pea rhizosphere [51]. Thus, GFP activity of the transformant is a useful tool for nondestructive monitoring of hyphal proliferation and competitive ability of the fungus in a natural soil, and such transformants may provide useful models to investigate the ecological fitness of introduced fungi in soil.

Sometimes, scanning confocal laser microscopy (SLCM, or "confocal") can provide several advantages over conventional light or standard epifluorescence microscopy for visualization of microbes, such as those in colonized plant debris in soil or inside target pathogen structures (e.g., sclerotia). A confocal microscope combines fluorescence microscopy with electronic image analysis to obtain three-dimensional images. Confocal microscopy has proven to be a powerful tool for examining the structure, organization, and physiology of microbial cells on surfaces, among other uses [50, 52, 53]. The shallow depth of field (as little as 0.5–1.5 μm) of a confocal microscope allows information to be collected from well-defined optical sections rather than from most of the specimen as in conventional microscopy. Thus, out-of-focus fluorescence is eliminated, resulting in increased contrast and clarity. Effectively, the sample can be optically sectioned, and stacks of optical sections taken at successive focal planes (i.e., a "Z-series") can be reconstructed to produce a focused view of the sample.

Spatial Complexity of Soil Biocontrol Systems

Biological organisms and their controlling variables rarely are distributed in a random or in a uniform way, since the environment is spatially structured by various energy inputs that result in patchy structures or gradients [54]. The rhizoplane is a good example of this, since energy input is largely due to root exudates, and certain zones of roots produce more exudate than others [55]. Although only 1-2% of a root system may be colonized by microbes, space can be a limiting factor [56, 57]. Sites may be preferentially colonized by some rhizoplane microbes and thus may no longer be available to others [58]. As a result, microbial distributions are not always random or uniform. The tendency for rhizosphere microbial population to conform to lognormal or similar frequency distributions has been noted [59–61], although there has been less attention to mechanisms of population development that might lead to such distributions.

Successful manipulation of rhizoplane microflora to enhance native or introduce beneficial microorganisms depends on knowledge of their ecological associations over time and space [62, 63]. Mandeel and Baker [64] observed that potential infection courts on the rhizoplane can be protected by an agent that actively competes for these sites and that this mechanism has more potential impact on biocontrol efficiency than does rhizosphere nutrient competition. They suggested that efficiency values are influenced by spatial relationships, especially relative proximities of the pathogen to penetration sites compared to proximity of the biocontrol agent. We would further add that these spatial arrangements and their interactions evolve over time, so that understanding them is critical to effective biocontrol. Additionally, most previous work in this area has been done using gnotobiotic systems (e.g., [65–68]), so effects of the indigenous microbial community on the distributions of the pathogen and the inoculated biocontrol agent are not yet well understood.

Spatial variability of exudates from seeds and roots may influence sites of colonization of biocontrol agents as well as pathogens. Chemotaxis of zoospores and growth of mycelium towards roots are highly regulated by root exudates [69–71]. Zoospores are an important infectious propagule for Pythium spp. The greatest accumulation of zoospores has been reported to be at approximately 2.5 mm behind the root tip (zone of cell elongation), where a major portion of diffusible compounds are exuded [71]. However, we found that Pythium ultimum var. sporangiiferum encysted throughout the root region and that the extent of zoospore aggregation was density dependent [66]. Pythium ultimum var. sporangiiferumrepresents a promising choice as a model organism to investigate spatial-temporal interactions between biocontrol agents and pathogens, since if access to spatially important entry sites for pathogens is blocked by biocontrol agents or other microbes, then infection and disease may be reduced. For example, Dandurand and Menge found that prior colonization of citrus roots by Fusarium solani reduced numbers of encysted zoospores of Phytophthora parasitica and P. citrophthora and reduced P. parasitica population [70, 72].

Deacon and Donaldson [73] described zoospores as "homing agents" or "site-selection agents," because their motility is linked to receptor functions for detecting environmental signals. They described the zoospore homing response as a sequence requiring two factors: a chemotactic stimulus and a suitable surface on which zoospores can orient [73]. Further, they pointed out that zoospores can precisely locate root tips, wounds, or even individual root cells, so that understanding the homing response and factors that may modify it is central to understanding zoosporic fungi and for attempts to control them. Spatial statistics may provide an effective tool for understanding this phenomenon and its ecological significance.

Geostatistical Analysis as a Tool to Evaluate Spatial Complexity

Nonpathogenic rhizoplane colonizers (biocontrol agents) at or near infection courts may be well positioned to modify the zoospore encystment process and subsequent root infection. For example, the frequency distribution of cucumber root sections without encysted zoospores ofPythium aphanidermatum was higher for roots treated with biocontrol bacteria than for untreated roots [74]. However, we suggest that frequency distribution analyses alone are not adequate to describe spatially structured biological phenomena [75–77], in part, because they assume independence of observations. The existence of spatial structure implies that the assumption of independence is not met, because any ecological phenomenon located at a given sampling point may have an influence on other points close by or some distance away [75].

What information can spatial statistics (e.g., geostatistics) provide that simple observation or distributional methods (e.g., frequency distributions and dispersion indices) do not? The variable success of biocontrol agents in controlling diseases may, in part, be caused by a lack of understanding of the spatial partitioning of resources in the rhizoplane. Spatial analysis of biocontrol agents may increase our predictive ability for effective biocontrol agents. For example, colonization patterns of bacteria were reported by Fukui et al. [20]; although it was observed that two strains colonized various parts of sugar beet seeds, a quantitative analysis of the spatial patterns of the two strains was not made, and conclusions derived from spatial patterns of the two bacteria based on observation only are difficult to interpret. A quantitative analysis of spatial patterns is particularly important to determine whether biocontrol agents change the spatial patterns of pathogens in predictable and consistent ways. Spatial statistical analysis provides a mechanism to explore processes that generate different patterns of organisms over time and helps to determine the sensitivity of spatial pattern to variations in these processes. Spatial analysis is defined here as any analysis that quantitatively evaluates variations or changes based on spatial orientation within a defined area or volume.

Frequency distribution methods are commonly based on mean/variance ratios, which do not provide reliable interpretations of spatial structure, since information on the location of each sample site is ignored. Although such indices are useful for estimation of population means, they do not maintain the spatial integrity of samples, making spatial analysis impossible [76–79]. Unlike frequency analysis, spatial analysis requires that the spatial integrity of observations be maintained; that is, spatial coordinates are recorded for each sample point. One method for spatial analysis, geostatistics, provides a quantitative assessment of spatial distributions that maintains the spatial integrity of data and is able to analyze the degree of association (autocorrelation) based on direction and distance between samples [80, 81]. Although geostatistics evolved primarily through mining applications, it has proven highly applicable to biological systems. For example, geostatistics has been used to evaluate insect spatial distributions [82, 83] and a spatial simulation model [77, 84], plant disease patterns [76,85], and spatial patterns of zoospores and bacteria on roots [66, 67, 86]. Some advantages of geostatistics may be summarized as follows: geostatistics is independent of the relationship between the mean and variance; geostatistics maintains the spatial integrity of locations of samples and uses the variation between points to evaluate spatial dependence; geostatistics assesses spatial dependence quantitatively and can be used to compare spatial dependence at different points in time or at the same point in time under different conditions.

Although geostatistical analysis cannot say specifically what the mechanism of biocontrol activity of T. harzianum is, it can help give credence to hypotheses about mechanisms. As an example, Mandeel and Baker [64] observed that potential infection courts on the rhizoplane can be protected by an agent that actively competes for these sites, and they suggested that this mechanism has more potential impact on biocontrol efficiency than does rhizosphere nutrient competition. Mandeel and Baker further observed that biocontrol efficiency values are influenced by spatial relationships, especially relative proximities of the pathogen to penetration sites compared to proximity of the biocontrol agent [64]. If space is indeed a mechanism in biocontrol efficacy, then proliferation of the biocontrol agent would be expected to generate a change in observed spatial patterns of the pathogen population, as our preliminary results have indicated. However, these preliminary results were obtained in a gnotobiotic system. Of course, a recurring question in biological control research is why is biocontrol efficacy so variable in natural soils? Again, spatial statistical analysis can provide quantitative answers about whether spatial attributes of a biocontrol agent will change or not change in the presence of indigenous microbes. Effective development of spatially rigorous analysis techniques will provide a necessary framework for evaluating the effects of mechanisms that are studied at the genetic and biochemical level.

Simulation Modeling of Biocontrol Agent Performance

Fungi in soil perform beneficial roles that include biological control of soil borne plant pathogens. However, relatively little predictive information is available about the growth and activity of fungal hyphae in soil habitats. A number of fungal growth models deal primarily with aspects of physiology and morphogenesis (e.g., [87, 88]), and these may not be directly applicable to predict fungal proliferation in natural habitats; other models deal with colony pattern formation in 2D and 3D space. Knudsen & Stack [89] introduced the idea of a simulation model for hyphal growth of a fungal hyperparasite through soil and use of the model to predict the incidence of hyperparasitism of sclerotia of certain soilborne plant pathogens. Lejeune and Baron [90] and Lejeune et al. [91] simulated the 3D growth of the filamentous fungus Trichoderma reesei, based on properties of mycelial growth (total hyphal length and total number of tips). Cross and Kenerley [92], using a combination of the Ratkowsky and Arrhenius equations, modeled colony growth of T. virens at different temperatures. Other models of fungal growth [93, 94] are useful for studying fungal physiology and morphogenesis but do not have an explicit spatial component (i.e., they assume spatial homogeneity) and thus may not be applicable to predict colony expansion in a heterogeneous environment.

Models developed for fungal growth in homogeneous environments, such as on agar surfaces or in culture media, may be less applicable to predict colony expansion in a heterogeneous environment. One potential way to deal with problems of spatial heterogeneity is the development of an "individual-based" model that tracks spatial attributes of the organism of interest as well as those of the local environment. With an individual-based model, information about individual organisms and/or habitat patches is retained in a database, and the model can then aggregate attributes of individuals to generate dynamics of the larger population. Thus, individual-based models represent an inductive approach to spatial population modeling in that population statistics are derived from collective attributes of individuals [95]. For example, we have used this approach to predict epizootics caused by the entomopathogen Beauveria bassiana in population of Russian wheat aphid on uniform versus nonuniform host plant distributions [77, 84].

There are important reasons why spatially homogeneous models of microbial growth kinetics and interactions are inadequate for some purposes, especially for analysis of mechanisms that generate different spatial dynamics as the population develops. If a biocontrol system is made more complex with patchily distributed agents and/or target pathogens, the spatial dimensions of the system become central to effective analysis and comprehensive understanding. Although description of spatially complex biological processes may sometimes most effectively be accomplished by modeling individuals, the concept of an "individual" is ambiguous for clonal organisms such as fungi. Numerically, physiologically, and often genetically, a single fungal colony is best described as a network of individual hyphal segments. Two advantages of this approach are as follows: first, the spatial structure of the fungal colony (hyphal network) can be explicitly reconstructed at any point in time; second, the approach has the flexibility to accommodate environments that are heterogeneous with respect to the spatial distribution of physical and biotic factors, and where phenotypic heterogeneity among individuals is present.

The objective of our "Fungmod" study [96] was to develop a modeling framework, in the form of an individual-based computer simulation model, that is suitable for exploration of the spatial dynamics of hyphal growth of a biocontrol fungus in a three-dimensional soil habitat. In that model, records of spatial location and branching hierarchy are maintained for individual hyphal nodes or for one or more fungal colonies. Thus, each hyphal segment is spatially referenced (coordinates in three dimensions) in terms of its ending node position, with its starting location referenced as the ending position of its parent node. In this way, the entire spatial structure of the fungal colony can be explicitly reconstructed from the data set at any point in time. The model proceeds at 1-day time steps and allows prediction of colony diameter, fungal biomass, and the spatial distribution of hyphae over time. The model was validated against data derived from image analysis of hyphal biomass accumulation in soil.

Although simulation models have been used extensively for arthropod biocontrol systems (e.g., [97]), relatively few models have been used for plant disease biocontrol studies. Knudsen and Hudler [98] described a simulation model to predict population dynamics of antibiotic-producing strains of Pseudomonas fluorescens on conifer foliage and subsequent effects on conidial germination of the fungal pathogen Gremmeniella abietina. Spurr and Knudsen [99] and Knudsen and Spurr [100] described how population models for antagonistic phylloplane bacteria might be incorporated into predictive models for foliar fungal diseases. However, those studies did not address spatial aspects of growth and proliferation of the putative control agents. For control of sclerotia-forming pathogens in soil, the ability of the biocontrol agent to effectively explore the three-dimensional soil space via hyphal growth probably is more important than spatially localized proliferation, for example, via sporulation. During the initial stage of colony development in soil, hyphae of biocontrol fungi will obtain their energy from the formulation matrix and/or from substrata located within the soil volume in which they are growing. The extent of mycoparasitism thus would depend on the density and spatial arrangement both of target sclerotia and of hyphae of the biocontrol agent. Quantification of patterns and pattern changes in hyphal growth should enhance prediction of the likelihood of successful mycoparasitic events.

Trophic Complexity Influences the Success of Fungal Biocontrol Agents

A wide range of biotic factors, in addition to abiotic factors (e.g., [101]), influence growth and proliferation of fungi introduced into natural soils. However, there has been relatively little investigation of quantitative effects of the indigenous soil biota on growth and establishment of fungi introduced into natural soils as potential biocontrol agents. Especially, interactions between introduced fungi and soil microanimals have largely been neglected in studies relating to the potential proliferation and ecological impact of introduced fungal agents. Nematodes are the most abundant soilinhabiting animals; there are many described species (more than 80,000), and population densities typically are high in the top 5 cm of soils, with substantial seasonal differences in agricultural soils [102, 103]. Identified nematode trophic groups include fungivores, bacterivores, omnivores, predators, and plant parasites [104]. Fungivorous nematodes have stylets and feed on mycelia of many different species of soil fungi, including beneficial fungi as well as plant pathogens [105].

Numerous species of fungivores have been found in soils [106]. The most common genera found in agricultural soils are Aphelenchoides, Aphelenchus, Tylenchus, and Ditylenchus [106,107]; some species within this group are plant parasites. Fungivores within this group of nematodes usually are found at lower densities than bacterivores or plant parasites [107]. However, if suitable fungi as food are available, population of fungivores may rapidly increase. Hofman and S'jacob [107] reported that numbers of the mycophagous nematodesAphelenchoides sp., Aphelenchus avenae, and unidentified microbivorous nematodes increased severalfold within a few days on dying roots of flax that had been precolonized by Rhizoctonia solani. Many plant ectoparasitic nematodes also may be fungivorous, as the need arises [106]. For example, Ditylenchus destructor, which is parasitic on carrot, potato, alfalfa, radish, and sweet potato, also can be cultured on a variety of filamentous fungi [107].

Fungivorous nematodes may be a significant biotic constraint on activity of biocontrol fungi in the field, and experiments suggest that their presence introduces a layer of trophic complexity that needs to be factored into attempts to predict the fitness of introduced biocontrol fungi in soil. In our laboratory, we isolated a fungivorous nematode (Aphelenchoides sp.) from field soil and found that population of the nematode significantly reduced hyphal growth and total population of T. harzianum under some environmental conditions, in nonsterile field soil [11]. Addition of Aphelenchoides resulted in lower numbers of sclerotia colonized by T. harzianumcompared to untreated controls [11]. Because the nematode feeds (via stylet) on hyphae of the fungus but is unable to feed on conidia or chlamydospores, active nematode population appears to have a disproportionate effect on fungal population that is actively growing (i.e., with a comparatively higher ratio of hyphae: spores) compared to population of Trichodermathat is primarily inactive. In a typical biocontrol scenario, the introduced fungus is applied to soil, often with a food base, from which it initiates hyphal growth (e.g., [10, 19, 34]). Meanwhile, the bulk of the indigenous Trichoderma population may be relatively inactive. Thus, if conditions are favorable for nematode activity, nematode feeding effectively will be selected against the introduced organism, and its apparent relative fitness compared to indigenous Trichoderma will be less. Perhaps ironically, soil environmental conditions that are favorable for nematode activity (adequate soil moisture and moderate temperature) are those that might otherwise be considered "ideal" for introduction of the fungal agent. The nematode x fungus interaction just described represents the first level of trophic complexity that may be a determinant of the (apparent) fitness of a recombinant fungus released into the environment.

Another aspect of trophic complexity may be implicated if hyphae of the biocontrol agent are able to contact and colonize the target substratum, that is, sclerotia of S. sclerotiorum, since colonized sclerotia potentially represent a refuge where the agent is protected from nematode activity and thus can persist. Similarly, because Trichoderma spp. are effective colonizers of plant debris, thecolonization of dead plant material may also provide a refuge from nematode activity. Refuge theory has played a prominent role in a large number of recent ecological studies, although to our knowledge it has not been addressed in microbial ecology. Hawkins et al. [108] applied principles of refuge theory to the biological control of insect pests using parasitoids and suggested that both theoretical and experimental results predict that success of biological control efforts (in insect pest/ parasitoid systems) is inversely related to the proportion of insects protected from parasitoid attack. In general, parasitism rates increased as refuge level decreased [109]. However, it was also pointed out that research on refuge theory in entomology has largely been restricted to correlative studies and theory and that theoretical work has advanced more rapidly than experimental tests of the models' assumptions in part due to the difficulty of manipulating the feeding biology and refuges of individual insects in natural systems [109]. In the Aphelenchoides x Trichoderma x Sclerotinia system, however, the roles are quite different: one hypothesized refuge for T. harzianum is sclerotia of S. sclerotiorum, the target organism itself. Ability of the biocontrol agent to rapidly colonize this refuge may simultaneously effectuate biocontrol of the target sclerotia and also escape from feeding activity of fungivorous nematodes (however, recent experiments suggest that the nematode may also be able to enter sclerotia and feed there; R. Garcia De la Cruz & G. R. Knudsen, unpublished). One aspect that these very different biological control systems (insect parasitoids and their hosts versus fungal hyperparasites and their hosts) have in common is, as described by Hawkins et al. [108], that refuges "...provide a general mechanism for interpreting ecological patterns at both the community level (their species diversity) and population level (their dynamics)." Again, it is anticipated that study of this phenomenon will be helpful in linking ecological theory with applied biotechnology risk assessment considerations.

CONCLUSIONS

As Spurr and Knudsen [99] noted, the search for a "silver bullet," a biological control organism that can simply be applied as if it were a chemical pesticide, is unlikely to succeed in an ecologically complex milieu. Ecological systems, as described by Levin [110], are the prototypical complex adaptive systems (CAS) "in which macroscopic system properties such as trophic structure, diversity-productivity relationships, and patterns of nutrient flux emerge from interactions among components, and may feed back to influence the subsequent development of those interactions." In recent years, the study of complex systems has contributed significantly to important research areas in ecology, namely, food (trophic) webs, patch dynamics and spatial structure, and population fluctuations [111–113]. We believe that each of these areas will continue to prove highly applicable and provide many research opportunities for biological control of soilborne plant pathogens.

REFERENCES

1. V. Vasas and F. Jordán, "Topological keystone species in ecological interaction networks: considering link quality and non-trophic effects," Ecological Modelling, vol. 196, no. 3-4, pp. 365–378, 2006.
2. E. A. Herre, N. Knowlton, U. G. Mueller, and S. A. Rehner, "The evolution of mutualisms: exploring the paths between conflict and cooperation," Trends in Ecology and Evolution, vol. 14, no. 2, pp. 49–53, 1999. · ·
3. C. Neuhauser and J. E. Fargione, "A mutualism-parasitism continuum model and its application to plant-mycorrhizae interactions," Ecological Modelling, vol. 177, no. 3-4, pp. 337–352, 2004. · ·
4. J. A. van Veen, L. S. van Overbeek, and J. D. van Elsas, "Fate and activity of microorganisms introduced into soil," Microbiology and Molecular Biology Reviews, vol. 61, no. 2, pp. 121–135, 1997.
5. J. E. Duffy, B. J. Cardinale, K. E. France, P. B. McIntyre, E. Thébault, and M. Loreau, "The functional role of biodiversity in ecosystems: incorporating trophic complexity,"Ecology Letters, vol. 10, no. 6, pp. 522–538, 2007. · ·
6. D. S. Srivastava and T. Bell, "Reducing horizontal and vertical diversity in a foodweb triggers extinctions and impacts functions," Ecology Letters, vol. 12, no. 10, pp. 1016–1028, 2009. · ·
7. J. S. Jacobs, R. L. Sheley, and B. D. Maxwell, "Effect of Sclerotinia sclerotiorum on the interference between bluebunch wheatgrass (Agropyron spicatum) and spotted knapweed (Centaurea maculosa)," Weed Technology, vol. 10, no. 1, pp. 13–21, 1996.

8. M. H. Abu-Dieyeh and A. K. Watson, "Efficacy of Sclerotinia minor for dandelion control: effect of dandelion accession, age and grass competition," Weed Research, vol. 47, no. 1, pp. 63–72, 2007. · ·

9. J. P. Hubbard, G. E. Harman, and Y. Hadar, "Effects of soilborne Pseudomonas spp. on the biological control agent, Trichoderma harzianum, on pea seeds," Phytopathology, vol. 73, pp. 655–659, 1983. ·

10. L. Bin, G. R. Knudsen, and D. J. Eschen, "Influence of an antagonistic strain ofPseudomonas fluorescens on growth and ability of Trichoderma harzianum to colonize sclerotia of Sclerotinia sclerotiorum in soil," Phytopathology, vol. 81, pp. 994–1000, 1991.

11. Y.-S. Bae and G. R. Knudsen, "Influence of a fungus-feeding nematode on growth and biocontrol efficacy of Trichoderma harzianum," Phytopathology, vol. 91, no. 3, pp. 301–306, 2001.

12. F. N. Martin and J. T. English, "Population genetics of soilborne fungal plant pathogens," Phytopathology, vol. 87, no. 4, pp. 446–447, 1997. ·

13. G. E. Harman, C. R. Howell, A. Viterbo, I. Chet, and M. Lorito, "Trichoderma species—opportunistic, avirulent plant symbionts," Nature Reviews Microbiology, vol. 2, no. 1, pp. 43–56, 2004. · ·

14. M. Vermaa, S. K. Brara, R. D. Tyagia, R. Y. Surampallib, and J. R. Valéroa, "Antagonistic fungi, Trichoderma spp.: panoply of biological control," Biochemical Engineering Journal, vol. 37, no. 1, pp. 1–20, 2007. ·

15. G. E. Harman, "Myths and dogmas of biocontrol: changes in perceptions derived from research on Trichoderma harzianum T-22," Plant Disease, vol. 84, no. 4, pp. 377–393, 2000.

16. M. C. Limón, M. R. Chacón, R. Mejías et al., "Increased antifungal and chitinase specific activities of Trichoderma harzianum CECT 2413 by addition of a cellulose binding domain," Applied Microbiology and Biotechnology, vol. 64, no. 5, pp. 675–685, 2004. · ·

17. M. R. Mousseaux, R. K. Dumroese, R. L. James, D. L. Wenny, and G. R. Knudsen, "Efficacy of Trichoderma harzianum as a biological control of Fusarium oxysporum in container-grown

Douglas-fir seedlings," New Forests, vol. 15, no. 1, pp. 11–21, 1998. · ·

18. L.-M. Dandurand, R. D. Mosher, and G. R. Knudsen, "Combined effects of Brassica napus seed meal and Trichoderma harzianum on two soilborne plant pathogens,"Canadian Journal of Microbiology, vol. 46, no. 11, pp. 1051–1057, 2000.
19. G. R. Knudsen, D. J. Eschen, L. M. Dandurand, and L. Bin, "Potential for control ofSclerotinia sclerotiorum through colonization of sclerotia by Trichoderma harzianum,"Plant Disease, vol. 75, pp. 466–470, 1991. ·
20. R. Fukui, E. I. Poinar, P. H. Bauer et al., "Spatial colonization patterns and interaction of bacteria on inoculated sugar beet seed," Phytopathology, vol. 84, no. 11, pp. 1338–1345, 1994.
21. G. E. Harman, R. Petzoldt, A. Comis, and J. Chen, "Interactions between Trichoderma harzianum strain T22 and maize inbred line Mo17 and effects of these interactions on diseases caused by Pythiuin ultimum and Colletotrichum graminicola," Phytopathology, vol. 94, no. 2, pp. 147–153, 2004.
22. W. Mao, R. D. Lumsden, J. A. Lewis, and P. K. Hebbar, "Seed treatment using pre-infiltration and biocontrol agents to reduce damping-off of corn caused by species ofPythium and Fusarium," Plant Disease, vol. 82, no. 3, pp. 294–299, 1998.
23. A. Muthukumar, A. Eswaran, S. Nakkeeran, and G. Sangeetha, "Efficacy of plant extracts and biocontrol agents against Pythium aphanidermatum inciting chilli damping-off,"Crop Protection, vol. 29, no. 12, pp. 1483–1488, 2010. · ·
24. T. C. Paulitz, "Effect of Pseudomonas putida on the stimulation of Pythium ultimum by seed volatiles of pea and soybean," Phytopathology, vol. 81, pp. 1282–1287, 1991. ·
25. G. E. Harman, "Myths and dogmas of biocontrol: changes in perceptions derived from research on Trichoderma harzianum T-22," Plant Disease, vol. 84, no. 4, pp. 377–393, 2000.
26. N. Benhamou and I. Chet, "Cellular and molecular mechanisms involved in the interaction between Trichoderma harzianum and Pythium ultimum," Applied and Environmental Microbiology, vol. 63, pp. 2095–2099, 1997.

27. C. R. Howell, "Mechanisms employed by Trichoderma species in the biological control of plant diseases: the history and evolution of current concepts," Plant Disease, vol. 87, no. 1, pp. 4–10, 2003.

28. A. Katayama and F. Matsumura, "Degradation of organochlorine pesticides, particularly endosulfan, by Trichoderma harzianum," Environmental Toxicology and Chemistry, vol. 12, no. 6, pp. 1059–1065, 1993.

29. O. Ermisch and H. J. Rehm, "Degradation of polycyclic aromatic hydrocarbons by the immobilized mold Trichoderma harzianum in soil," in Proceedings of the DECHEMA Annual Meeting on Biotechnology and 58th Meeting of the European Biotechnology Federations, and Joint Meeting of Society for Industry Microbiology (SIM ‹89), pp. 780–781, 1989.

30. I. Grishkan, E. Nevo, S. P. Wasser, and A. Beharav, "Adaptive spatiotemporal distribution of soil microfungi in "Evolution Canyon" II, Lower Nahal Keziv, western Upper Galilee, Israel," Biological Journal of the Linnean Society, vol. 78, no. 4, pp. 527–539, 2003. · ·

31. H. Green and D. F. Jensen, "A tool for monitoring Trichoderma harzianum: II. The use of a GUS transformant for ecological studies in the rhizosphere," Phytopathology, vol. 85, no. 11, pp. 1436–1440, 1995.

32. C.-L. Zhang, I. S. Druzhinina, C. P. Kubicek, and T. Xu, "Trichoderma biodiversity in China: evidence for a North to South distribution of species in East Asia," FEMS Microbiology Letters, vol. 251, no. 2, pp. 251–257, 2005. · ·

33. G. R. Knudsen and L. Bin, "Effects of temperature, soil moisture and wheat bran on growth of Trichoderma harzianum from alginate pellets," Phytopathology, vol. 80, pp. 724–727, 1990. ·

34. G. R. Knudsen, D. J. Eschen, L. M. Dandurand, and Z. G. Wang, "Method to enhance growth and sporulation of pelletized biocontrol fungi," Applied and Environmental Microbiology, vol. 57, no. 10, pp. 2864–2867, 1991.

35. R. D. Lumsden, J. P. Carter, J. M. Whipps, and J. M. Lynch, "Comparison of biomass and viable propagule measurements in the antagonism of Trichoderma harzianumagainst Pythium

ultimum," Soil Biology and Biochemistry, vol. 22, no. 2, pp. 187–194, 1990.

36. J. Rigot and F. Matsumura, "Assessment of the rhizosphere competency and pentachlorophenol-metabolizing activity of a pesticide-degrading strain of Trichoderma harzianum introduced into the root zone of corn seedlings," Journal of Environmental Science and Health B: Pesticides, Food Contaminants, and Agricultural Wastes, vol. 37, no. 3, pp. 201–210, 2002. · ·
37. C.-T. Lo, E. B. Nelson, C. K. Hayes, and G. E. Harman, "Ecological studies of transformed Trichoderma harzianum strain 1295-22 in the rhizosphere and on the phylloplane of creeping bentgrass," Phytopathology, vol. 88, no. 2, pp. 129–136, 1998.
38. I. N. Roberts, R. P. Oliver, P. J. Punt, and C. A. M. J. J. Van den Hondel, "Expression of the Escherichia coli -glucuronidase gene in industrial and phytopathogenic filamentous fungi," Current Genetics, vol. 15, no. 3, pp. 177–180, 1989.
39. C. Thrane, M. Lubeck, H. Green et al., "A tool for monitoring Trichoderma harzianum: I. Transformation with the GUS gene by protoplast technology," Phytopathology, vol. 85, no. 11, pp. 1428–1435, 1995.
40. M. Chalfie, Y. Tu, G. Euskirchen, W. W. Ward, and D. C. Prasher, "Green fluorescent protein as a marker for gene expression," Science, vol. 263, no. 5148, pp. 802–805, 1994.
41. B. P. Cormack, G. Bertram, M. Egerton, N. A. R. Gow, S. Falkow, and A. J. P. Brown, "Yeast-enhanced green fluorescent protein (yGFP): a reporter of gene expression inCandida albicans," Microbiology, vol. 143, no. 2, pp. 303–311, 1997. ·
42. J. Sheen, H. S. Hwang Seongbin, Y. Niwa, H. Kobayashi, and D. W. Galbraith, "Green-fluorescent protein as a new vital marker in plant cells," Plant Journal, vol. 8, no. 5, pp. 777–784, 1995. · ·
43. T. Spellig, A. Bottin, and R. Kahmann, "Green fluorescent protein (GFP) as a new vital marker in the phytopathogenic fungus Ustilago maydis," Molecular and General Genetics, vol. 252, no. 5, pp. 503–509, 1996.
44. A. J. Vanden Wymelenberg, D. Cullen, R. N. Spear, B. Schoenike, and J. H. Andrews, "Expression of green fluorescent protein in

Aureobasidium pullulans and quantification of the fungus on leaf surfaces," Bioechniques, vol. 23, no. 4, pp. 686–690, 1997.

45. Y. S. Bae and G. R. Knudsen, "Cotransformation of Trichoderma harzianum with -glucuronidase and green fluorescent protein genes provides a useful tool for monitoring fungal growth and activity in natural soils," Applied and Environmental Microbiology, vol. 66, no. 2, pp. 810–815, 2000. · ·

46. Y. S. Bae and G. R. Knudsen, "Soil microbial biomass influence on growth and biocontrol efficacy of Trichoderma harzianum," Biological Control, vol. 32, no. 2, pp. 236–242, 2005. · ·

47. K. A. Orr and G. R. Knudsen, "Use of green fluorescent protein and image analysis to quantify proliferation of Trichoderma harzianum in nonsterile soil," Phytopathology, vol. 94, no. 12, pp. 1383–1389, 2004.

48. T. G. Kim and G. R. Knudsen, "Quantitative real-time PCR effectively detects and quantifies colonization of sclerotia of Sclerotinia sclerotiorum by Trichoderma spp,"Applied Soil Ecology, vol. 40, no. 1, pp. 100–108, 2008. · ·

49. T. G. Kim and G. R. Knudsen, "Colonization of Sclerotinia sclerotiorum sclerotia by a biocontrol isolate of Trichoderma harzianum, and effects on myceliogenic germination,"Biocontrol Science and Technology, vol. 19, no. 10, pp. 1081–1085, 2009. · ·

50. T. G. Kim and G. R. Knudsen, "Comparison of real-time PCR and microscopy to evaluate sclerotial colonisation by a biocontrol fungus," Fungal Biology, vol. 115, no. 4-5, pp. 317–325, 2011. · ·

51. T. G. Kim and G. R. Knudsen, "Relationship between the biocontrol fungusTrichoderma harzianum and the phytopathogenic fungus Fusarium solani f.sp. pisi,"Applied Soil Ecology, vol. 68, pp. 57–60, 2013. ·

52. C.-T. Huang, F. P. Yu, G. A. McFeters, and P. S. Stewart, "Nonuniform spatial patterns of respiratory activity within biofilms during disinfection," Applied and Environmental Microbiology, vol. 61, no. 6, pp. 2252–2256, 1995.

53. J. S. McLean, O. N. Ona, and P. D. Majors, "Correlated biofilm imaging, transport and metabolism measurements via combined

nuclear magnetic resonance and confocal microscopy," ISME Journal, vol. 2, no. 2, pp. 121–131, 2008. · ·

54. P. Legendre and M. J. Fortin, "Spatial pattern and ecological analysis," Vegetatio, vol. 80, no. 2, pp. 107–138, 1989. · ·
55. A. D. Rovira, "Zones of exudation along plant roots and spatial distribution of microoorganisms in the rhizosphere," Pesticide Science, vol. 4, no. 3, pp. 361–366, 1973. ·
56. S. Compant, C. Clément, and A. Sessitsch, "Plant growth-promoting bacteria in the rhizo- and endosphere of plants: their role, colonization, mechanisms involved and prospects for utilization," Soil Biology and Biochemistry, vol. 42, no. 5, pp. 669–678, 2010. · ·
57. A. Ramette and J. M. Tiedje, "Multiscale responses of microbial life to spatial distance and environmental heterogeneity in a patchy ecosystem," Proceedings of the National Academy of Sciences of the United States of America, vol. 104, no. 8, pp. 2761–2766, 2007. · ·
58. A. J. H. Davey, C. P. Doncaster, and O. D. Jones, "Distinguishing between interference and exploitation competition for shelter in a mobile fish population," Environmental Modeling and Assessment, vol. 14, no. 5, pp. 555–562, 2009. · ·
59. E. Limpert, W. A. Stahel, and M. Abbt, "Log-normal distributions across the sciences: keys and clues," BioScience, vol. 51, no. 5, pp. 341–352, 2001.
60. J. E. Loper, T. V. Suslow, and M. N. Schroth, "Lognormal distribution of bacterial populations in the rhizosphere," Phytopathology, vol. 74, pp. 1454–1460, 1984. ·
61. G. Wieland, R. Neumann, and H. Backhaus, "Variation of microbial communities in soil, rhizosphere, and rhizoplane in response to crop species, soil type, and crop development," Applied and Environmental Microbiology, vol. 67, no. 12, pp. 5849–5854, 2001. · ·
62. J. M. Raaijmakers, T. C. Paulitz, C. Steinberg, C. Alabouvette, and Y. Moënne-Loccoz, "The rhizosphere: a playground and battlefield for soilborne pathogens and beneficial microorganisms," Plant and Soil, vol. 321, no. 1-2, pp. 341–361, 2009. · ·

63. M. E. Stanghellini and S. L. Rasmussen, "Root prints: a technique for the determination of the in situ spatial distribution of bacteria on the rhizoplane of field-grown plants,"Phytopathology, vol. 79, pp. 1131–1134, 1989. ·
64. Q. Mandeel and R. Baker, "Mechanisms involved in biological control of Fusarium wilt on cucumber with strains of non-pathogenic Fusarium oxysporum," Phytopathology, vol. 81, pp. 462–469, 1991. ·
65. T. F. C. Chin A Woeng, W. de Priester, A. J. van der Bij, and B. J. Lugtenberg, "Description of the colonization of a gnotobiotic tomato rhizosphere by Pseudomonas fluorescens biocontrol strain WCS365, using scanning electron microscopy," Molecular Plant Microbe Interactions, vol. 10, no. 1, pp. 79–86, 1997. ·
66. L. M. Dandurand, G. R. Knudsen, and D. J. Schotzko, "Quantification of Pythium ultimum var. sporangiiferum zoospore encystment patterns using geostatistics,"Phytopathology, vol. 85, no. 2, pp. 186–190, 1995.
67. L. M. Dandurand, D. J. Schotzko, and G. R. Knudsen, "Spatial patterns of rhizoplane populations of Pseudomonas fluorescens," Applied and Environmental Microbiology, vol. 63, pp. 3211–3217, 1997.
68. M. Simons, A. J. van der Bij, I. Brand, L. A. de Weger, C. A. Wijffelman, and B. J. J. Lugtenberg, "Gnotobiotic system for studying rhizosphere colonization by plant growth-promoting Pseudomonas bacteria," Molecular Plant Microbe Interactions, vol. 9, pp. 600–607, 1996. ·
69. C. A. Walker and P. van West, "Zoospore development in the oomycetes," Fungal Biology Reviews, vol. 21, no. 1, pp. 10–18, 2007. · ·
70. L. M. Dandurand and J. A. Menge, "Influence of Fusarium solani on chemotaxis of zoospores of Phytophthora parasitica and Phytophthora citrophthora and on distribution of 14C in citrus tissues and root exudate," Soil Biology and Biochemistry, vol. 26, no. 1, pp. 75–79, 1994.
71. R. T. Mitchell and J. W. Deacon, "Differential (host-specific) accumulation of zoospores of Pythium on roots of graminaceous and non-graminaceous plants," New Phytologist, vol. 102, pp. 113–122, 1986.

72. L. M. Dandurand and J. A. Menge, "Influence of Fusarium solani on citrus root rot caused by Phytophthora parasitica and Phytophthora citrophthora," Plant and Soil, vol. 144, no. 1, pp. 13–21, 1992. · ·

73. J. W. Deacon and S. P. Donaldson, "Molecular recognition in the homing responses of zoosporic fungi, with special reference to Pythium and Phytophthora," Mycological Research, vol. 97, no. 10, pp. 1153–1171, 1993. ·

74. T. Zhou and T. C. Paulitz, "In vitro and in vivo effects of Pseudomonas spp. on Pythium aphanidermatum: zoospore behavior in exudates and on the rhizoplane of bacteria-treated cucumber roots," Phytopathology, vol. 83, pp. 872–876, 1993. ·

75. P. Legendre, "Spatial autocorrelation: trouble or new paradigm?" Ecology, vol. 74, no. 6, pp. 1659–1673, 1993.

76. P. C. Nicot, D. I. Rouse, and B. S. Yandell, "Comparison of statistical methods for studying spatial patterns of soilborne plant pathogens in the field," Phytopathology, vol. 74, pp. 1399–1402, 1984.

77. D. J. Schotzko and G. R. Knudsen, "Use of geostatistics to evaluate a spatial simulation of Russian wheat aphid (Homoptera: Aphididae) movement behavior on preferred and nonpreferred hosts," Environmental Entomology, vol. 21, no. 6, pp. 1271–1282, 1992.

78. P. A. Jumars, D. Thistle, and M. L. Jones, "Detecting two-dimensional spatial structure in biological data," Oecologia, vol. 28, no. 2, pp. 109–123, 1977. · ·

79. A. J. Sawyer, "Inconstancy of Taylor›s b: simulated sampling with different quadrat sizes and spatial distributions," Researches on Population Ecology, vol. 31, no. 1, pp. 11–24, 1989. · ·

80. E. H. Isaaks and R. H. Srivastava, Applied Geostatistics, Oxford University Press, Oxford, UK, 1989.

81. B. B. Trangmar, R. S. Yost, and G. Uehara, "Application of geostatistics to spatial studies of soil properties," Advances in Agronomy, vol. 38, pp. 45–94, 1985.

82. W. P. Kemp, T. M. Kalaris, and W. F. Quimby, "Rangeland grasshopper (Orthoptera: Acrididae) spatial variability: macroscale

population assessment," Journal of Economic Entomology, vol. 82, pp. 1270–1276, 1989.

83. D. J. Schotzko and C. M. Smith, "Effect of host plant on the plant to plant distribution of the Russian wheat aphid (Homoptera: Aphididae)," Journal of Economic Entomology, vol. 84, pp. 1725–1734, 1991.

84. G. R. Knudsen and D. J. Schotzko, "Spatial simulation of epizootics caused by Beauveria bassiana in Russian wheat aphid populations," Biological Control, vol. 16, no. 3, pp. 318–326, 1999. · ·

85. D. A. Johnson, J. R. Alldredge, J. R. Allen, and R. Allwine, "Spatial pattern of downy mildew in hop yards during severe and mild diesease epidemics," Phytopathology, vol. 81, pp. 1369–1374, 1991. ·

86. G. R. Knudsen and L. M. Dandurand, "Analysis of spatial patterns of rhizoplane colonization," in The Spatial Distribution of Microbes in the Environment, R. B. Franklin and A. L. Mills, Eds., pp. 109–133, Springer, New York, NY, USA, 2007.

87. L. Edelstein, "The propagation of fungal colonies: a model for tissue growth," Journal of Theoretical Biology, vol. 98, no. 4, pp. 679–701, 1982.

88. L. Edelstein, Y. Hadar, and I. Chet, "A model for fungal colony growth applied toSclerotium rolfsii," Journal of General Microbiology, vol. 129, no. 6, pp. 1873–1881, 1983.

89. G. R. Knudsen and J. P. Stack, "Modeling growth and dispersal of fungi in natural environments," in Handbook of Applied Mycology, Vol. I: Soil and Plants, D. K. Arora, K. G. Mukerji, B. Rai, and G. R. Knudsen, Eds., pp. 625–645, Marcel Dekker, New York, NY, USA, 1991.

90. R. Lejeune and G. V. Baron, "Simulation of growth of a flamentous fungus in 3 dimensions," Biotechnology and Bioengineering, vol. 53, no. 2, pp. 139–150, 1997.

91. R. Lejeune, J. Nielsen, and G. V. Baron, "Morphology of Trichoderma reesei QM 9414 in submerged cultures," Biotechnology and Bioengineering, vol. 47, no. 5, pp. 609–615, 1995. · ·

92. D. Cross and C. M. Kenerley, "Modelling the growth of Trichoderma virens with limited sampling of digital images," Journal of Applied Microbiology, vol. 97, no. 3, pp. 486–494, 2004. · ·

93. A. P. J. Trinci, "Regulation of hyphal branching and hyphal orientation," in The Ecology and Physiology of Fungal Mycelium, D. H. Jennings and A. D. M. Rayner, Eds., pp. 23–52, Cambridge University Press, Cambridge, UK, 1984.

94. J. G. H. Wessels, "A steady-state model for apical wall growth in fungi," Acta Botanica Neerlandica, vol. 37, pp. 3–16, 1988.

95. M. Huston, D. DeAngelis, and W. Post, "New computer models unify ecological theory," BioScience, vol. 38, pp. 682–691, 1988. ·

96. G. R. Knudsen, J. P. Stack, S. O. Schuhmann, K. Orr, and C. LaPaglia, "Individual-based approach to modeling hyphal growth of a biocontrol fungus in soil," Phytopathology, vol. 96, no. 10, pp. 1108–1115, 2006. · ·

97. A. A. de Souza, S. G. F. Martins, and M. S. Zacarias, "Computer simulation applied to the biological control of the insect Aphis gossypii for the parasitoid Lysiphlebus testaceipes," Ecological Modelling, vol. 220, no. 6, pp. 756–763, 2009. · ·

98. G. R. Knudsen and G. W. Hudler, "Use of a computer simulation model to evaluate a plant disease biocontrol agent," Ecological Modelling, vol. 35, no. 1-2, pp. 45–62, 1987.

99. H. W. Spurr Jr. and G. R. Knudsen, "Biological control of leaf diseases with bacteria," inBiological Control on the Phylloplane, C. E. Windels and S. E. Lindow, Eds., pp. 45–62, American Phytopathological Society, St. Paul, Minn, USA, 1985.

100. G. R. Knudsen and H. W. Spurr Jr., "Field persistence and efficacy of five bacterial preparations for control of peanut leaf spot," Plant Disease, vol. 71, no. 5, pp. 442–445, 1987.

101. B. K. Duffy, B. H. Ownley, and D. M. Weller, "Soil chemical and physical properties associated with suppression of take-all of wheat by Trichoderma koningii,"Phytopathology, vol. 87, no. 11, pp. 1118–1124, 1997.

102. E. A. Paul and F. E. Clark, "Components of the soil biota," in Soil Microbiology and Biochemistry, pp. 69–107, Academic Press, New York, NY, USA, 2nd edition, 1996.

103. G. W. Yeates, D. A. Wardle, and R. N. Watson, "Responses of soil nematode populations, community structure, diversity and temporal variability to agricultural intensification over a seven-year period," Soil Biology and Biochemistry, vol. 31, no. 12, pp. 1721–1733, 1999. · ·

104. W. L. Nicholas, The Biology of Free-Living Nematodes, Clarendon, Oxford, UK, 2nd edition, 1984.

105. G. L. Barnes, C. C. Russell, W. D. Foster, and R. W. McNew, "Aphelenchus avenae, a potential biological control agent for root rot fungi," Plant Disease, vol. 65, pp. 423–424, 1981. ·

106. D. W. Freckman and E. P. Caswell, "The ecology of nematodes in agroecosystems,"Annual Review of Phytopathology, vol. 23, pp. 275–296, 1985. ·

107. T. W. Hofman and J. J. S›jacob, "Distribution and dynamics of mycophagous and microbivorous nematodes in potato fields and their relationship to some food sources,"Annals of Applied Biology, vol. 115, no. 2, pp. 291–298, 1989.

108. B. A. Hawkins, M. B. Thomas, and M. E. Hochberg, "Refuge theory and biological control," Science, vol. 262, no. 5138, pp. 1429–1432, 1993.

109. E. E. Porter and B. A. Hawkins, "Coexistence of specialist parasitoids with host refuges in the laboratory and the dynamics of spatial heterogeneity in attack rate," Oikos, vol. 100, no. 2, pp. 232–240, 2003. · ·

110. S. A. Levin, "Ecosystems and the biosphere as complex adaptive systems," Ecosystems, vol. 1, no. 5, pp. 431–436, 1998.

111. M. Anand, A. Gonzalez, F. Guichard, J. Kolasa, and L. Parrott, "Ecological systems as complex systems: challenges for an emerging science," Diversity, vol. 2, no. 3, pp. 395–410, 2010.

112. J. Jabiol, B. G. McKie, A. Bruder, C. Bernadet, M. O. Gessner, and E. Chauvet, "Trophic complexity enhances ecosystem functioning in an aquatic detritus-based model system,"Journal of Animal Ecology, vol. 82, no. 5, pp. 1042–1051, 2013

113. R. E. Ulanowicz, R. D. Holt, and M. Barfield, "Limits on ecosystem trophic complexity: insights from ecological network analysis," Ecology Letters, vol. 17, no. 2, pp. 127–136, 2014.

Chapter

5

Myco-Biocontrol of Insect Pests: Factors Involved, Mechanism, and Regulation

Sardul Singh Sandhu[1], Anil K. Sharma[1], Vikas Beniwal[1], Gunjan Goel[1], Priya Batra1, Anil Kumar[2], Sundeep Jaglan[3], A. K. Sharma[4], and Sonal Malhotra[1]

[1]Department of Biotechnology, Maharishi Markandeshwar, Mullana, Ambala, Haryana 133203, India

[2]Department of Bio & Nanotechnology, Guru Jambheshwar University of Science & Technology, Hisar, Haryana 125001, India

[3]Plant Biotechnology Division, Indian Institute of Integrative Medicine (IIIM), Jammu, J & K 180001, India

[4]Department of Pulmonary Medicine & Critical Care, Mayo Clinic, Rochester, MN 55905, USA

ABSTRACT

The growing demand for reducing chemical inputs in agriculture and increased resistance to insecticides have provided great impetus to the development of alternative forms of insect-pest control. Myco-biocontrol offers an attractive alternative to the use of chemical pesticides. Myco-biocontrol agents are naturally occurring organisms which are perceived as less damaging to the environment. Their mode of action appears little complex which makes it highly unlikely that resistance could be developed to a biopesticide. Past research has shown some promise of the use of fungi as a selective pesticide. The current paper updates us about the recent progress in the field of myco-biocontrol of insect pests and their possible mechanism of action to further enhance our understanding about the biological control of insect pests.

INTRODUCTION

Myco-biocontrol is an environmentally sound and effective means of reducing or mitigating insect-pests and its effects through the use of natural enemies. Pest-related damages result in a heavy loss, approximately estimated to be US $10,000 million annually in agricultural production in the field and storage in India. Myco-biocontrol is the use of fungi in biological processes to lower the insect density with the aim of reducing disease-producing activity and consequently crop damage [1, 2]. All groups of insects may be affected and over 700 species of fungi have been recorded as pathogens. Some of these fungi have restricted host ranges, for example, Aschersonia aleyrodes infects only scale insects and whiteflies, while other fungal species have a wide host range, with individual isolates being more specific to target pests. Some species are facultative generalist pathogens, such as Aspergillus and Fusarium. However, most species are obligate pathogens, often quite specific and rarely found, for example, many species of Cordyceps.

Entomopathogens such as M. anisopliae and B. bassiana are well characterized in respect to pathogenicity to several insects and have been used as myco-biocontrol agents for biological control of agriculture pests worldwide. About 11 companies offer at least 16 products based on the entomopathogenic fungiB. bassiana at Columbia. These products are not only used in coffee crop but also in other crops such as bean, cabbage, corn, potato, and tomato. They are used to treat haematophagous insect pests and vectors of diseases like mosquitoes and flies. Under natural conditions, fungi are the frequent and often important natural mortality factor in insect populations.

Unlike other potential biocontrol agents, fungi do not have to be ingested to infect their hosts but invade directly through the cuticle, and so can, potentially, be used for control of all insects including sucking insects.

SOURCES OF MYCO-BIOCONTROL AGENTS

A substantial number of mycoinsecticides and mycoacaricides have been developed worldwide since the 1960s. Products based on Beauveria bassiana(33.9%) [3], Metarhizium anisopliae (33.9%), Isaria fumosorosea (5.8%), and B. brongniartii (4.1%) are the most common among the 171 products [4]. Approximately 75% of all listed products are currently registered, undergoing registration or commercially available, whereas 15% are no longer available. Insects in the orders Hemiptera, Coleoptera, Lepidoptera, Thysanoptera, and Orthoptera comprise most of the targets. Research has focused on the relatively easily produced asexual spores (conidia) of the hyphomycete generaMetarhizium [5], Beauveria, Verticillium, and Paecilomyces. These fungi have a wide host range although there is considerable genetic diversity within species and some clades show a high degree of specificity. For example, Metarhizium anisopliae var. acridum [6] is only effective against acridid insects (grasshoppers and locusts). The commercial Beauveria bassiana-based mycoinsecticides are relatively stable compared with other biological insect control agents for lepidopteran insect pests [7, 8].

Major concern about the ill effects of the chemicals pesticides along with advances in biotechnology has promoted search for new and ecofriendly insect control alternatives. Among them biological control is one of the most effective alternative. Biological control can be pronounced to involve the use of biological entities for reducing the damage caused by insect pest population. Very less percentage of known insect species are considered as pests, and the control of these insects has been a major challenge before scientific community since the beginning of the agriculture era. Although effective, chemical pesticides are expensive and provide only temporary relief, as the explosive reproductive and evolutionary capacities of the insect allow them to develop mechanisms resistance to these and other control strategies. Apart from the dangers posed upon human population especially health, they also affect nontarget organism and cause irreversible damage to the environment by disturbing the ecological balance. For biocontrol to become an integral part of the modern agriculture, a few goals must be met such as the selection and development of superior biocontrol agents, the development of fermentation system for biomass production, and the development of formulation and delivery systems, which are compatible with microorganism requirements as well as with common agriculture practices.

Past researches have shown fungi being a potential biological control agent mainly due to their high reproductive capabilities, target specific activity, short generation time, and resting stage or saprobic phase-producing capabilities that can ensure their survival for a longer time when no host is present. Primary requirement for the use of an entomogenous fungus as a myco-biocontrol agent is the susceptibility of the insect on one hand and virulence of the fungus on the other hand. The latter depends on selection of a strain with stable, specific efficacy for a target host. Hence there is an immense potential for genetic improvement of fungi for myco-biocontrol.

Deuteromycetes fungi has a broad host range, and in particular Metarhiziumand Beauveria show particular promise as myco-biocontrol agents and are currently being used as myco-insecticides. Beauveria bassiana andMetarhizium anisopliae are among the first entomopathogenic fungi being successfully used for the myco-biocontrol of insect pests. Molecular techniques for genetic engineering for filamentous fungi provide new opportunities for the study of fungi used in myco-biocontrol of insect pests. The isolation of gene encoding pathogenesis and virulence allows rigorous testing of their role in pathogenesis and should provide a rational basis for strain improvement.

The studies on the fungal pathogenesis for myco-biocontrol of insect pests are still at the preliminary stages. However, the development of molecular biological technique for entomopathogenic fungi such as Beauveria bassianaand Metarhizium anisopliae, which are coupled with cloning of genes encoding putative pathogenesis determinants, will create more potential candidates to manage the notorious insect pests population. Research in this field will inevitably depend on the development if sensitive techniques for monitoring the environmental fate of recombinant strain for management of insect pests are designed.

ENTOMOPATHOGENIC FUNGI

Entomopathogenic fungi are among the first organisms to be used for the biological control of pests. More than 700 species of these fungi from around 90 genera are pathogenic to insects. Most of them are found within the deuteromycetes and entomophthorales. Some entomopathogenic fungi have restricted host ranges, for example, Aschersonia aleyrodes infects only scale insects, and whiteflies, while other fungal species have a wide host range, with individual isolates being more specific to target pests. Entomopathogens such as M. anisopliae and B. bassiana are well characterized in respect to pathogenicity to several insects, and they have been used as agents for the biological control of agriculture pests worldwide. About 11 companies offer at least 16 products based on the entomopathogenic fungi B. bassiana. These products are not only used in the coffee crop but also in other crops such as bean, cabbage, corn, potato, and tomato. They are also used to treat haematophagous insect pests and vectors of diseases like mosquitoes and flies [9]. Biopesticide has very complex mode of action unlike chemical pesticides, therefore resistance in pest could not be developed.

BIO-MANAGEMENT OF INSECT-PESTS BY ENTOMOPATHOGENIC FUNGI

The entomogenous word has been derived from two Greek words, "entomon" meaning insects and "genes" meaning arising in. Therefore, the etymological meaning of entomogenous microorganism is "microorganisms which arise in insects." The power of these entomogenous microorganisms in bringing about a certain degree of natural or microbial control of insect pests is directly related to human welfare which has attracted the attention of microbiologist, molecular biologists, and entomologists in the recent years. Several entomopathogens, when inundatively introduced into a variety of habitats, can provide effective long-term to short-term control. Table 1 gives an overview of different types of organisms, including pathogenic microorganisms such as viruses, bacteria, fungi, protozoa, and nematodes for their use as biocontrol agents. The most propitious integration of pathogens, predators, insect's growth regulators, and conventional insecticides may provide us with long-term control of serious agricultural insect pests. Some of the potential candidates for myco-biocontrol of insect pests are discussed here.

Table 1: Different types of organisms including pathogenic microorganisms such as viruses, bacteria, fungi, protozoa, and nematodes have been used as biocontrol agents been used as biocontrol agents

Biocontrol agent	Common examples	Biological action
Parasitic insects: living organisms remaining in close association with their hosts and gradually derive their food from the host	Trichogramma chilonis, Epiricania melanoleuca	They live and feed internally or externally on the host.
Predatory insects		
	Chrysoperla carnea, Cryptolaemus montrouzieri	Insects which kill and devour the prey
Microorganisms (Bacteria, fungi, viruses)		
	Bacteria, for example, Bacillus thuringiensis Fungi, for example, species of Trichoderma, Nomuraea, Paecilomyces, Verticillium, Metarhizium, and Beauveria Viruses, for example, nuclear polyhedrosis virus	Cause diseases in pests and inhibit the harmful fungi

Beauveria sp

Beauveria bassiana, a filamentous fungus, belongs to a class of insect pathogenic deuteromycete also known as imperfect fungus. Strains ofBeauveria are highly adapted to particular host insects. A broad range of B. bassiana spp. have been isolated from a variety of insect worldwide which are of medicinal or agricultural importance. Beauveria bassiana is a fungus that grows naturally in soils throughout the world and acts as a pathogen on various insect species, causing white muscardine disease, therefore belongs to the entomopathogenic fungi [10–14]. An interesting feature of Beauveria sp. is the high host specificity of many isolates. Hosts of medicinal importance include vectors for agents of tropical infectious diseases such as tsetse flyGlossina morsitans, and sand fly Phlebotomus that transmits Leishmania and bugs of genera Triatoma and Rhodnius, the vectors of Chagas disease. Hosts of agricultural and forest significance include the Colorado potato beetle, the codling moth, and several genera of termites, American bollworm Helicoverpa armigera [15], Hyblaeapara and Eutectona machaeralis. Furthermore, the high level of persistence in the host population and in the environment provides long-term effects of the entomopathogenic fungi on pest suppression, if an epizootic is caused. It is being used as a biological insecticide to control a number of pests such as termites, whitefly, and in malaria-transmitting mosquitoes [16, 17]. B. bassiana is the anamorph (asexually reproducing form) of Cordyceps bassiana. The latter teleomorph (the sexually reproducing form) has been collected only in eastern Asia [3]. Rehner and Buckley [18] have shown that B. bassiana consists of many distinct lineages that should be recognized as distinct phylogenetic species. This ubiquitous fungus has long been known to be the most common causative agent of disease associated with dead and moribund insects in nature [19] and has been scrutinized worldwide as a microbial control agent of hypogeous species [20]. Many curculionidae weevils with a sub-terranean larval stage are highly susceptible to this white muscardine disease [21]. Like many species of entomogenous fungi, B. bassiana is composed of many genetically distinct variants associated with geographical location and host which differ substantially in their ability to produce pathogenesis. As an insecticide, the spores are sprayed on affected crops as an emulsified suspension or wettable powder. B. bassiana parasitizes a very wide range of arthropod hosts and therefore is considered as a nonselective biological insecticide. B. bassiana is also applied against the European corn borer Ostrinia Mubilalis, pine caterpillars Dendrolimus spp., and green leafhoppers Nephotettix spp.

Verticillium lecanii

Another entomopathogenic fungus Verticillium lecanii is a widely distributed fungus, which can cause large epizootic in tropical and subtropical regions, as well as in warm and humid environments [22]. It was reported by Kim et al. [23] that V. lecanii was an effective biological control agent againstTrialeurodes vaporariorum in South Korean greenhouses. This fungus attacks nymphs and adults and stucks to the leaf underside by means of a filamentous mycelium [22]. In 1970s, Verticillium lecanii was developed to control whitefly and several aphids species, including the green peach aphids (Myzus persicae) for use in the greenhouse chrysanthemums [16].

Verticillium lecanii was considered as a major parasite which caused a massive decline of cereal-cyst nematode populations in monocultures of susceptible crops [24]. Verticillium chlamydosporium has a wide host range amongst cyst and root-knot nematodes but it is very variable and only some isolates may have potential as commercial biological control agents.

Metarhizium spp

Metarhizium anisopliae is also very potential pathogen on insect pests and is explored for myco-biocontrol of notorious insect pests [10, 25]. A complete bioactivity of M. anisopliae has been tested on teak skeletonizer Eutectona machaeralis and found M. anisopliae to be a potential myco-biocontrol agent of teak pest [26]. Hasan et al. [27] have tested spore production of M. anisopliae by solid state fermentation.

Nomuraea sp

Nomuraea rileyi another potential entomopathogenic fungi is a dimorphic hyphomycete that can cause epizootic death in various insects. It has been shown that many insect species belonging to Lepidoptera including Spodoptera litura and some belonging to

Coleoptera are susceptible to N. rileyi [28]. The host specificity of N. rileyi and its ecofriendly nature encourage its use in insect pest management. Although, its mode of infection and development have been reported for several insect hosts such as Trichoplusia ni, Heliothis zea,Plathypena scabra, Bombyx mori, Pseudoplusia includes, and Anticarsia gemmatalis. Another insect Spilosoma was found to be severely attacked byNomuraea rileyi, hence studied in detail for its myco-biocontrol [29]. Similarly an epizootic of Nomuraea rileyi was observed on Junonia orithya [30] which was proved to be the best alternative to manage the hedge plant eater Junonia orithya.

Paecilomyces sp

Paecilomyces is a genus of nematophagous fungus which kills harmful nematodes by pathogenesis, causing disease in the nematodes. Thus, the fungus can be used as a bionematicide to control nematodes by applying to soil (Table 2). Paecilomyces lilacinus principally infects and assimilates eggs of root-knot and cyst nematodes. The fungus has been the subject of considerable biological control research following its discovery as a biological control agent in 1979. Paecilomyces fumosoroseus (Wize) Brown and Smith [31] (Hyphomycetes) is one of the most important natural enemies of whiteflies worldwide, and causes the sickness called "Yellow Muscardine" [22]. Strong epizootic potential against Bemisia and Trialeurodes spp. in both greenhouse and open field environments has been reported. P. lilacinus has been considered to have the greatest potential for application as a biocontrol agent in subtropical and tropical agricultural soils. The ability of this fungus to grow extensively over the leaf surface under humid conditions is a characteristic that certainly enhances its ability to spread rapidly through whitefly populations [32].

Table 2: Various bioactive products derived from entomopathogenic fungi which could be commercially useful for field application have been illustrated

Product	Fungus	Biological action
Mycotal	*Verticillium lecanii*	Fungal pesticide
Pfr21	*Paecilomyces fumosoroseus*	Fungal pesticide
Verelac	*Verticillium lecanii*	Sucking pests
Beevicide	*Beauveria bassiana*	Borer type pests
Grubkill	Selected fungus and bacteria	Borers and sucking pest
Pelicide	*Paecilomyces lilacinus*	Effective against nematode
Biologic Bio 1020	*Metarhizium anisopliae*	Mycelium granules as pesticide
Bioter	*Verticillium lecanii*	Effective against termites
Brocaril	*Beauveria bassiana*	Wettable powder used as pesticide
Ostrinil	*Beauveria bassiana*	Microgranules of mycelium used as pesticide
Boverol	*Beauveria bassiana*	Dry pellets as pesticide
Naturalis	*Beaveria bassiana*	Liquid formulation as pesticide
Mycontrol-WP	*Beauveria bassiana*	Wettable powder as pesticide
Betel	*Beauveria brongniartii*	Microgranules of mycelium used as pesticide
Engerlingspilz	*Beauveria brongniartii*	Barley kernels colonized with fungus used as pesticide
Biopath	*Metarhizium anisopliae*	Conidia on a medium used as pesticides
Biomite	*Verticillium lecanii* and other entomopathogenic organisms	Effective against mites
Biogreen	*Metarhizium anisopliae*	Conidia produced on grain used as pesticide
Naturalis-O and BotaniGard	*Beauveria bassiana*	Effective against whiteflies

Trypae Mix	*Trichoderma* and *Paecilomyces*	Effective against fungal pathogens and nematodes in soil

Natural epizootics of these fungi suppress Bemisia tabaci populations. Epizootics caused by Paecilomyces fumosoroseus also lead to substantially reductions in B. tabaci populations during or immediately following rainy seasons or even prolonged periods of cool, humid conditions in the field or greenhouse [4]. However, in general, epizootics of naturally occurring fungi cannot be relied upon for control. Only a few species of fungi have the capacity to cause high level of mortality, and development of natural epizootics which is not only dependent on the environmental conditions, but also influenced by various crop production practices. Also, epizootics often occur after intense injury has already been inflicted by whiteflies [4]. Kim et al. [23] reported thatP. fumosoroseus is best for controlling the nymphs of whitefly. These fungi cover the whitefly's body with mycelial threads and stick them to the underside of the leaves. The nymphs show a "feathery" aspect and are surrounded by mycelia and conidia [22]. P. furiosus is also used to control mosquito sp. Culex pipiens [25].

MODE OF ACTION OF ENTOMOPATHOGENIC FUNGUS

Entomopathogenic fungi constitute the largest single group of insect pathogens among microorganisms. Such insect killing fungi are very fast Microorganisms to be recognized as disease causing agent in insects. Entomogenous fungi are promising myco-biocontrolling agent for a number of crop pests. Several species belonging to order Lepidoptera, Coleoptera, Homoptera, Hymenoptera, and Diptera are susceptible to various fungal infections. Entomopathogenic fungi have a great potential as myco-biocontrol agents, as they constitute a group with over 750 species that, when dispersed in the environment, provoke fungal infections in insect populations.

THE INFECTION PROCESS

Fungi have an unique mode of infection; they reach the haemocoel through the cuticle or possibly through the mouth parts. Ingested fungal spores do not germinate in the gut and are voided in the faeces. The death of the insect results from a combination of factors: mechanical damage resulting from tissue invasion, depletion of nutrient resources and toxicosis, and production of toxin in the body of insect.

Conidial Attachment with the Cuticle

For most of the entomopathogenic fungi host location is a random event and attachment is a passive process with the aid of wind or water. Attachment of a fungal spore to the cuticle surface of a susceptible host represents the initial event in the establishment of mycosis. It was observed that dry spores of B. bassiana possess an outer layer composed of interwoven fascicles of hydrophobic rodlets. This rodlet layer appears to be special to the conidial stage and has not been reported on the vegetative cells. The adhesion of dry spores to the cuticle was suggested to be due to nonspecific hydrophobic forces imposed by the rodlets [33]. Some of these moieties like lectins, a kind of carbohydrate binding glycoproteins, have also been detected on the conidial surface of B. bassiana. It was also observed that lectins could be involved in binding between conidia and the insect cuticle. The exact mechanisms responsible for the interaction between fungal spores and the cuticle remain to be determined [34]. When the pathogen reaches and adheres to the host surface, it proceeds with rapid germination and growth which are profoundly influenced by the availability of water, nutrients, oxygen as well as pH, and temperature, and by the effects of toxic host-surface compound. Fungi with a broad host range germinate in culture in response to a wide range of nonspecific carbon and nitrogen sources [35]. Entomopathogenic fungi with restricted host range appear to have more specific requirements for germination [36].

Formation of an Infection Structure

Entomopathogenic fungi invade their hosts by infection process: penetration of the host cuticle or put pressure on cuticle by making appressorium and then penetrate by penetration peg [35]. The cuticle has two layers: the outer epicuticle and the procuticle. The epicuticle is a very complex thin structure that lacks chitin but contains phenol-stabilized proteins and is covered by a waxy layer containing fatty acids, lipids and sterols [37]. The procuticle forms the majority of the cuticle and contains chitin fibrils embedded into a protein matrix together with lipids and quinones [38]. Protein may account for up to 70% of the cuticle. In many areas of the cuticle, the chitin is organized helically giving rise to a laminate structure. Entomopathogenic fungi, B. bassianaconidia germinate on the host surface and differentiate an infection structure termed appressorium. The appressorium represents an adaptation for concentrating physical and chemical energy over a very small area so that access may be achieved efficiently (Figure 1). Thus, formation of the appressorium plays a pivotal role in establishing a pathogenic interaction with the host. Appressorium formation may be influenced by host surface topography, and biochemical investigations indicate the involvement of the intracellular second messengers Ca^{2+} and cyclic AMP (cAMP) in appressorium formation [39] or in general when the cuticle in hard [35].

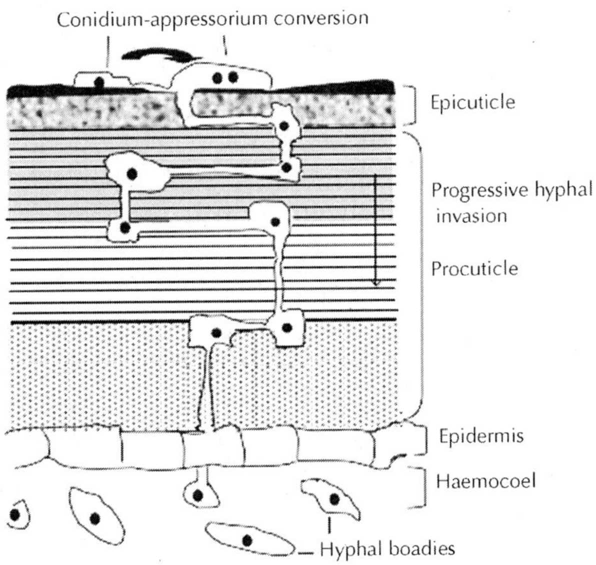

Figure 1: Depiction of the infection process in B. bassiana: structure of the insect cuticle and mode of penetration of fungal hyphae. Formation of the ap-

pressorium from the conidia helps in cuticle invasion and subsequent hyphal penetration to the haemocoel. Tissue invasion by hyphae and proliferation of hyphal bodies leads to insect death.

Penetration of the Cuticle

Entomopathogenic fungi need to penetrate through the cuticle into the insect body to obtain nutrients for their growth and reproduction. Entry into the host involves both enzymic degradation and mechanical pressure as evidenced by the physical separation of lamellae by penetrated hyphae. A range of extracellular enzymes that can degrade the major components of insect cuticle, including chitinases, lipases, esterases and at least four different classes of proteases, have been suggested to function during the fungal pathogenesis. Although the complex structure of the insect cuticle suggests that penetration would require the synergistic action of several different enzymes, much of the attention has focused on the cuticle-active endoprotease as a key factor in the process.

The production of cuticle-degrading enzymes by M. anisopliae during infection structure formation on Calliphora vomitoria and Manduca sexta has been investigated by biochemical and histochemical analyses both in vivo andin vitro. Among the first enzymes produced on the cuticle are endoproteases (termed PR1 and PR2) and aminopeptidases, coincident with the formation of appressoria. N-Acetylglucosaminidase is produced at a slow rate as compared to the proteolytic enzymes [40].

These fungi begin their infective process when spores are retained on the integument surface, where the formation of the germinative tube initiates, the fungi starts excreting enzymes such as proteases, chitinases, quitobiases, lipases, and lipoxygenases. V. lecanii is capable of penetrating the insect cuticle only with its germ tube while M. anisopliae and B. bassiana produce specific infection hyphae originating at appressoria. After the successful penetration, the fungus is then distributed into the haemolymph by formation of blastospores [41].

Different works are going on all over the world to distinguish the various enzymes which are required for the mechanism of entomopathogenicMetarhizium anisopliae, M. flaviviridae,

Paecilomyces farinosus, Beauveria bassiana, and B. brongniartii. Host specificity may be associated with the physiological state of the host system (i.e., insect maturation and host plant) [42], the properties of the insect's integument with the nutritional requirements of the fungus [43], and the cellular defense of the host [44]. In contrast to bacteria and viruses that pass through the gut wall from contaminated food, fungi have a unique mode of infection. They reach the haemocoel through the cuticle.

PRODUCTION OF TOXINS

A plethora of work with circumstantial evidence is available from deuteromycete pathogens for the involvement of fungal toxins in host death. The action of cytotoxins is suggested by cellular disruption prior to hyphae penetration. Behavioural symptoms such as partial or general paralysis, sluggishness, and decreased irritability in mycosed insects are consistent with the action of neuromuscular toxins [45]. B. bassiana and M. anisopliaeproduced significant amounts of toxic compounds within their hosts. For example, the toxins Beauvericin, Bassianolide, Isarolides, and Beauverolides have been isolated from B. bassiana infected hosts [46, 47], toxins Destruxins (DTXs) and Cytochalasins have been isolated from M. anisopliae infected hosts. The toxins have shown to have diverse effects on various insect tissues. DTX depolarizes the lepidopteran muscle membrane by activating calcium channels. In addition, function of insect hemocytes can be inhibited by DTX [48]. Presumably, there are still many toxins that remain to be isolated from parasitized insects and except DTXs, their relevance in the process of pathogenicity remains to be studied in detail.

GENETIC ENGINEERING STUDIES OF ENTOMOPATHOGENIC FUNGI

A more widespread use of fungi for myco-biocontrol will depend on improvements of wild-type strains by combining characteristics of different strains and mutants. Two types of improvements could be considered: (i) improving the efficacy of the insecticide, by reducing the dose necessary to kill the insects, by reducing the time to kill the pest or decreasing crop damage caused by the pest by reducing the feeding time; (ii) expanding the host range. Essential for the development of a hypervirulent strain is a complete understanding of the remarkable pathology of fungal infections. Molecular biology provides the necessary tools for dissecting the mechanisms of pathogenesis and in the longer term for producing recombinant organisms with new and relevant characteristics. Initial development towards these goals has occurred with M. anisopliae and to a much lesser extent with B. bassiana[49]. Genetic transformation systems, which are an essential part of modern fungal research, and are necessary for the experimental manipulation of virulence genes in vitro and in vivo, have been established [27, 50, 51]. The success of utilizing these procedures depends on the availability of selectable transformation markers [26]. Transformation techniques have been used to isolate specific pathogenic genes, investigate virulence determinants of M. anisopliae and B. bassiana, and to produce a strain with enhanced virulence. Unravelling the molecular mechanisms of fungal pathogenesis in insects will provide the basis for the genetic engineering of entomopathogenic fungi.

MOLECULAR STUDIES OF ENTOMOPATHOGENIC FUNGI

Implementation of PCR-based tools for characterization of organisms has greatly advanced the understanding of the phylogenies and species in entomopathogenic fungi, especially in B. bassiana and M. anisopliae. These fungi have received a lot of interest due to their potential as biocontrol agents of pests. A number of unspecific DNA-based methods have been used specially in Beauveria [52]. Random amplified polymorphic DNA (RAPD) has been used in many studies. It is based on the use of short general primers that anneal to unspecified regions in the template DNA whereas universally primed (UP) PCR is based on longer general primers and a higher annealing temperature which makes it more robust in terms of reproducibility [53–56]. UP-PCR has been used to separate sympatric isolates of Beauveria in Denmark and was used to place isolates in genetic groups [57]. For ecological studies, random amplified polymorphic DNA (RAPD) was used in combination with specific methods to separate isolate genotypes of M. anisopliae from Canadian soil [58] and to relate these genotypes to the origin of isolation. Thakur et al. [13] studied forty-eight isolates of indigenous strains of B. bassiana collected from Central India employing protease zymography and RAPD analysis. High genetic and biochemical diversities were indicated with a clear group of strains from Lepidopteran and Coleopteran insect hosts.

Different strategies have also been used for the analysis-RFLPs (restriction Fragment Length Polymorphism), AFLPs (amplified fragment length polymorphism). Digestion of PCR products of specific DNA regions, such as genes or ITS (internal transcribed spacer), with restriction enzymes, yields fragments of variable sizes. These RFLPs have been used for the characterization of both Beauveria and Metarhizium species [58]. B. bassianaand M. anisopliae [59–61] have been characterized using AFLP (amplified fragment length polymorphism), inter-simple-sequence repeats (ISSR), simple sequence repeats (SSRs), or microsatellites. Internal transcribed spacer sequences (ITSs) have been widely used in fungal systematics [6, 62]. In the case of the genus Paecilomyces, the analysis of sequences of the large and the small subunit rRNA gene has already indicated the polyphyly of the genus [63,64]. Recent development of microsatellite markers [65, 66] has surely provided an insight in the population ecology of B. bassiana and M. anisopliae. B. bassiana has been linked to plants as an endophytic fungus [67], and M. anisopliae has been shown to be associated with the rhizosphere of plants [68]. EST (expressed sequence tag) analysis of entomopathogenic fungus Beauveria(Cordyceps) bassiana has been studied using cDNA libraries [69]. EST analyses of two subspecies of M. anisopliae revealed distinct patterns of expression of proteases and pathogenicity factors. These expression patterns have led to the ability to examine gene expression during infection of various insect hosts [70,71].

MYCO-BIOCONTROL AGENTS

The advantages of using fungi as myco-biocontrol agents are as follows (1) Their high degree of specificity for pest control. Fungi can be used to control harmful insect pests without affecting beneficial insect predators and nonharmful parasites. (2) The absence of effects on mammals and thus the reduction of the hazards normally encountered with insecticide applications, such as pollution of the environment. (3) The lack of problems caused to insect resistance and prolonged pest control. (4) A high potential for further development by biotechnological research. (5) High persistence in the environment provides long-term effects of entomopathogenic fungi on pest suppression. However, there are also a number of constraints on the use of fungi as insecticides: (1) 2-3 weeks are required to kill the insects whereas chemical insecticides may need only 2-3 hours. (2) Application needs to coincide with high relative humidity, low pest numbers, and a fungicide free period (3) Due to the high specificity additional control agents are needed for other pests. (4) Their production is relatively expensive and the short shelf life of spores necessitates cold storage. (5) The persistence and efficacy of entomopathogenic fungi in the host population varies among different insects species, thus insect-specific application techniques need to be optimised to retain long-term impacts. (6) A potential risk to immunosuppressive people.

Entomopathogenic fungi are important as they are virulent, infect by contact, and persist in environment for a long period of time. These can be mass produced in liquid or solid media. Most of the entomopathogenic fungi are facultative parasites which exist as saprotrophs and therefore can be grown apart from living hosts. Few groups are obligate parasites which must be reared in living hosts. Introduction of fungal pathogens into the host population initiates epizootic and prevents or reduces damage by the pest. The initiation of artificial epizootics has been accomplished for long-term control especially in areas where high humidity condition prevails. There are several defense mechanisms in insect which prevents the penetration and the growth of the fungus. The most common is the melanisation of the cuticle at infection site. Entomopathogenic fungi can display either a very broad host spectrum like M. anisopliae, B. bassiana or have a very narrow host range likeAschersonia spp. [41].

CONCLUSIONS

Modern techniques in genetic engineering and biotechnology are extremely helpful in manipulating the desired traits in entomopathogenic fungi which can further improve its bioactivity. Numerous advantages one can foresee of using these fungal pathogens as pest control agents. They are host-specific having a wide host range and more importantly being less toxic to animals. Biological control agents have shown a lot of promise in terms of activity, though its efficacy is affected by many factors such as biotic and nonbiotic factors, host plant, and at the level of nematode infestation. There is a strong urge to elucidate the essence of these factors to improve the overall efficacy of these control agents along with developing novel methods to deliver sufficient inoculum at the target sites. Modern techniques in biotechnology has the potential to manipulate desirable traits of these entomopathogenic fungi to improve the overall field activity.

REFERENCES

1. I. Chet, "Trichoderma application, mode of action and potential as biocontrol agent of soil borne plant pathogenic fungi," in Innovative Approaches Plant Disease Control, I. Chet, Ed., pp. 137–160, John Wiley & Sons, New York, NY, USA, 1987.
2. T. Chet, H. Schichler, S. Haran, and A. B. Appenheim, "Cloned chitinase and their role in biological control of plant pathogenic fungi," inProceedings of the International Symposium on Chitin Enzymology, pp. 47–48, Senigalia, Italy, 1993.
3. Z. Z. Li, C. R. Li, B. Huang, and M. Z. Meizhen, "Discovery and demonstration of the teleomorph of Beauveria bassiana (Bals.) Vuill., an important entomogenous fungus," Chinese Science Bulletin, vol. 46, no. 9, pp. 751–753, 2001.
4. M. de Faria and S. P. Wraight, "Biological control of Bemisia tabaci with fungi," Crop Protection, vol. 20, no. 9, pp. 767–778, 2001.
5. G. Hu and J. St. Leger, "Field studies using a recombinant mycoinsecticide (Metarhizium anisopliae) reveal that it is rhizosphere competent," Applied and Environmental Microbiology, vol. 68, no. 12, pp. 6383–6387, 2002.
6. F. Driver, R. J. Milner, and J. W. H. Trueman, "A taxonomic revision ofMetarhizium based on a phylogenetic analysis of rDNA sequence data,"Mycological Research, vol. 104, no. 2, pp. 134–150, 2000.

7. I. S. Rana, A. Kanojiya, and S. S. Sandhu, "Mosquito larvicidal potential of fungi isolated from larval mosquito habitats against Aedes aegypti,"Journal of Biological Control, vol. 22, pp. 179–183, 2008.
8. R. Thakur, N. Jain, R. Pathak, and S. S. Sandhu, "Practices in wound healing studies of plants," Evidence-Based Complementary and Alternative Medicine, vol. 2011, Article ID 438056, 17 pages, 2011.
9. F. J. P. Florez, "Fungi for coffee berry borer control—Colombia," inProceedings of the 35th Annual Meeting of the Society of Invertebrate Pathology, Foz Do Iguassu, Brazil, 2002.
10. S. S. Sandhu, R. C. Rajak, and G. P. Agarwal, "Studies on prolonged storage of Beauveria bassiana conidia: effects of temperature and relative humidity on conidial viability and virulence against chickpea borer Helicoverpa armigera," Biocontrol Science & Technology, vol. 3, pp. 47–53, 1993.
11. S. S. Sandhu, S. E. Unkles, R. C. Rajak, and J. R. Kinghorn, "Generation of benomyl resistant Beauveria bassiana strains and their infectivity against Helicoverpa armigera," Biocontrol Science and Technology, vol. 11, no. 2, pp. 245–250, 2001.
12. R. Thakur, R. C. Rajak, and S. S. Sandhu, "Biochemical and molecular characteristics of indigenous strains of the entomopathogenic fungusBeauveria bassiana of Central India," Biocontrol Science and Technology, vol. 15, no. 7, pp. 733–744, 2005.
13. S. S. Sandhu and P. Vikrant, "Myco-insecticides: control of insect pests," in Microbial diversity: Opportunities & Challenges, S. P. Gautam, S. S. Sandhu, A. Sharma, and A. K. Pandey, Eds., Indica Publishers, New Delhi, India, 2004.
14. N. Jain, I. S. Rana, A. Kanojiya, and S. S. Sandhu, "Characterization ofBeaveria bassiana strains based on protease and lipase activity and their role in pathogenecity," Journal of Basic & Applied Mycology, vol. I-II, pp. 18–22, 2008.
15. R. Thakur and S. S. Sandhu, "Distribution, occurrence and natural invertebrate hosts of indigenous entomopathogenic fungi of Central India," Indian Journal of Microbiology, vol. 50, no. 1, pp. 89–96, 2010.

16. R. A. Hamlen, "Biological control of insects and mites on European greenhouse crops: research and commercial implementation,"Proceedings of the Florida State Horticultural Society, vol. 92, pp. 367–.368, 1979.
17. G. McNeil Donald Jr., "Fungus fatal to mosquito may aid global war on malaria," The New York Times, 2005.
18. S. A. Rehner and E. Buckley, "A Beauveria phylogeny inferred from nuclear ITS and EF1-α sequences: evidence for cryptic diversification and links to Cordyceps teleomorphs," Mycologia, vol. 97, no. 1, pp. 84–98, 2005.
19. D. M. Mcleod, "Investigations on the genera Beauveria Vuill. andTritirachium Limber," Canadian Journal of Botany, vol. 32, pp. 818–890, 1954.
20. P. Ferron, "Pest control by the fungi Beauveria and Metarhizium," inMicrobial Control of Insects and Mites, H. D. Burgess, Ed., pp. 465–482, Academic Press, New York, NY, USA, 1981.
21. J. B. Beavers, C. W. McCoy, and D. T. Kaplan, ""Natural enemies of sub-terranean Diaprepes abbbreviatus (Coleoptera: Curculionidae) larvae in Florida," Environmental Entomology, vol. 12, pp. 840–843, 1983.
22. E Nunez, J. Iannacone, and H. Gómez, "Effect of two entomopathogenic fungi in controlling aleurodicus cocois (Curtis, 1846) (Hemiptera: Aleyrodidae)," Chilean Journal of Agricultural Research, vol. 68, no. 1, pp. 21–30, 2008.
23. J. J. Kim, M. H. Lee, C. S. Yoon, H. S. Kim, J. K. Yoo, and K. C. Kim, "Control of cotton aphid and greenhouse whitefly with a fungal pathoge," Journal of National Institute of Agricultural Science and Technolog, pp. 7–14, 2002.
24. B. R Kerry, D. Crumph, and A. Mullen, "Studies of the cereal-cyst nematode, Heterodera avenue under continuous cereals, 1975–1978. II. Fungal parasitism of nematode females and eggs," Annals of Applied Biology, vol. 100, pp. 489–499, 1982.
25. S. S. Sandhu and M. Mishra, "Larvicidal activity of fungal isolatesBeaveria bassiana, Metarhizium anisopliae and Aspergillus flavus against mosquito sp. Culex pipiens," in Proceedings of the National Symposium on Advances in Biological Control of Insect Pests, pp. 145–150, Muzaffarnagar, India, 1994.

26. S. S. Sandhu, R. C. Rajak, and S. K. Hasija, "Potential of entomopathogens for the Biological management of medically important pest: progress and prospect," in Glimpses in Plant Sciences, vol. 2000, pp. 110–117, 2000.
27. S. Hasan, A. K. Bhamra, K. Sil, R. C. Rajak, and S. S. Sandhu, "Spore production of Metarhizium anisopliae (ENT-12) By Solid State fermentation," Journal of Indian Botanical Society, vol. 8, pp. 85–88, 2002.
28. C. M. Ignoffo, "The fungus Nomuraea rileyi as a microbial insecticide: fungi," in Microbial Control of Pests and Plant Diseases, H. D. Burges, Ed., pp. 513–538, Academic Press, London, UK, 1981.
29. S. O. Mathew, S. S. Sandhu, and R. C. Rajak, "Bioactivity of Nomuraea rileyi against Spilosoma obliqua: effect of dosage, temperature and relative humidity," Journal of Indian Botanical Society, vol. 77, pp. 23–25, 1998.
30. R. C. Rajak, S. S. Sandhu, S. Mukherjee, S. Kekre, and A. Gupta, "Natural outbreak of Nomuraea rileyi on Junonia orithyia," Journal of Biological Control, vol. 5, no. 2, pp. 123–124, 1991.
31. H. Seryczyńska and C. Bajan, "Defensive reactions of L3, L4 larvae of the Colorado beetle to the insecticidal fungi Paecilomyces farinosus (Dicks) Brown et Smith, Paecilomyces fumoso-roseus (Wize), Beauveria bassiana (Bols/Vuill.) (Fungi Imperfecti: Moniliales)," Bulletin de l›Academie Polonaise des Sciences. Serie des Sciences Biologiques, vol. 23, no. 4, pp. 267–271, 1975.
32. S. P. Wraight, R. I. Carruthers, S. T. Jaronski, C. A. Bradley, C. J. Garza, and S. Galaini-Wraight, "Evaluation of the entomopathogenic fungiBeauveria bassiana and Paecilomyces fumosoroseus for microbial control of the silverleaf whitefly, Bemisia argentifolii," Biological Control, vol. 17, no. 3, pp. 203–217, 2000.
33. D. G. Boucias, et al., "Nonspecific factors involved in the attachment of entomopathogenic deuteromycetes to host insect cuticle," Applied and Environment Microbiology, vol. 54, no. 7, pp. 1795–1805, 1988.
34. J. P. Latge, M. Monsigny, and M. C. Prevost, "Visualization of exocellular lectins in the entomopathogenic fungus Conidiobolus

obscurus," Journal of Histochemistry and Cytochemistry, vol. 36, no. 11, pp. 1419–1424, 1988.

35. S. S. Sandhu, "Effect of physical factors on germination of entomopathogenic fungus Beauveria bassiana conidia," National Academy of Science Letters, vol. 18, no. 1-2, pp. 1–5, 1995.

36. R. J. S. Leger, T. M. Butt, M. S. Goettel, R. C. Staples, and D. W. Roberts, "Productionin vitro of appressoria by the entomopathogenic fungusMetarhizium anisopliae," Experimental Mycology, vol. 13, no. 3, pp. 274–288, 1989.

37. R. H. Hackman, "Cuticle: biochemistry," in Biology of the Integument, J. Bereiter-Hahn, A. G. Matolts, and K. S. Richards, Eds., pp. 626–637, Springer, Berlin, Germany, 1984.

38. A. C. Neville, "Cuticle: organisation," in Biology of the Integument, J. Bereiter-Hahn, A. G. Matolts, and K. S. Richards, Eds., pp. 611–625, Springer, Berlin, Germany, 1984.

39. R. J. S. Leger, D. W. Roberts, and R. C. Staples, "A model to explain differentiation of appressoria by germlings of Metarhizium anisopliae,"Journal of Invertebrate Pathology, vol. 57, no. 3, pp. 299–310, 1991.

40. R. J. S. Leger, T. M. Butt, R. C. Staples, and D. W. Roberts, "Synthesis of proteins including a cuticle-degrading protease during differentiation of the entomopathogenic fungus Metarhizium anisopliae," Experimental Mycology, vol. 13, no. 3, pp. 253–262, 1989.

41. A. Bhattacharyya, A. C. Samal, and S. Kar, "Entomophagous fungus in pest management," News Letter, vol. 5, p. 12, 2004.

42. G. W. McCoy, R. A. Samson, and D. G. Boucias, "Entomogenous fungi," in CRC Handbook of Natural Pesticides, Part A. Entomogenous Protozoa and Fungi, C. M. Ignoffo, Ed., CRC Press, 1988.

43. J. L. Kerwin and R. K. Washino, "Oosporogenesis by Lagenidium giganteum: induction and maturation are regulated by calcium and calmodulin," Canadian Journal of Microbiology, vol. 32, no. 8, pp. 663–672, 1986.

44. M. M. Amer, T. I. El-Sayed, H. K. Bakheit, S. A. Moustafa, and Y. A. El-Sayed, "Pathogenicity and genetic variability of five entomopathogenic fungi against Spodoptera littoralis," Research

Journal of Agriculture and Biological Sciences, vol. 4, no. 5, pp. 354–367, 2008.

45. R. J. S. Leger, A. K. Charnley, and R. M. Cooper, "Characterization of cuticle-degrading proteases produced by the entomopathogenMetarhizium anisopliae," Archives of Biochemistry and Biophysics, vol. 253, no. 1, pp. 221–232, 1987.
46. R. L. Hamill, H. R. Sullivan, and M. Gorman, "Determination of pyrrolnitrin and derivatives by gas-liquid chromatography," Applied microbiology, vol. 18, no. 3, pp. 310–312, 1969.
47. J. F. Elsworth and J. F. Grove, "Cyclodepsipeptides from Beauveria bassiana bals. Part 1. Beauverol-ides H and I," Journal of the Chemical Society, Perkin Transactions 1, no. 3, pp. 270–273, 1977.
48. G. A. Bradfisch and S. L. Harmer, "omega-conotoxin GVIA and nifedipine inhibit the depolarizing action of the fungal metabolite, destruxin B on muscle from the tobacco budworm (Heliothis virescens)," Toxicon, vol. 28, no. 11, pp. 1249–1254, 1990.
49. D. D. Hegedus, T. A. Pfeifer, J. M. MacPherson, and G. G. Khachatourians, "Cloning and analysis of five mitochondrial tRNA-encoding genes from the fungus Beauveria bassiana," Gene, vol. 109, no. 1, pp. 149–154, 1991.
50. M. S. Goettel, R. J. S. Leger, S. Bhairi et al., "Pathogenicity and growth ofMetarhizium anisopliae stably transformed to benomyl resistance,"Current Genetics, vol. 17, no. 2, pp. 129–132, 1990.
51. S. S. Sandhu and P. Vikrant, "Evaluation of mosquito larvicidal toxins in the extra cellular metabolites of two fungal genera Beauveria andTrichoderma," in Emerging Trends in Mycology, Plant Pathology and Microbial Biotechnology, G. Bagyanarayana, B. Bhadraiah, and I. K. Kunwar, Eds., BS, Hyderabad, India, 2006.
52. T. R. Glare, S. D. Reay, T. L. Nelson, and R. Moore, "Beauveria caledonica is a naturally occurring pathogen of forest beetles,"Mycological Research, vol. 112, no. 3, pp. 352–360, 2008.
53. S. A. Bulat, M. Lübeck, N. Mironenko, D. F. Jensen, and P. S. Lübeck, "UP-PCR analysis and ITS1 ribotyping of strains of Trichoderma andGliocladium," Mycological Research, vol. 102, no. 8, pp. 933–943, 1998.

54. S. A. Bulat, M. Lubeck, I. A. Alekhina, D. F. Jensen, I. M. B. Knudsen, and P. S. Lubeck, "Identification of a universally primed-PCR-derived sequence-characterized amplified region marker for an antagonistic strain of Clonostachys rosea and development of a strain-specific PCR detection assay," Applied and Environmental Microbiology, vol. 66, no. 11, pp. 4758–4763, 2000.

55. M. Lübeck, I. A. Alekhina, P. S. Lübecks, D. F. Jensen, and S. A. Bulat, "Delineation of Trichoderma harziannum into two different genotypic groups by a highly robust fingerprinting method, UP-PCR, and UP-PCR product cross-hybridization," Mycological Research, vol. 103, no. 3, pp. 289–298, 1999.

56. T. K. Sabu, R. T. Shiju, K. Vinod, and S. Nithya, "A comparison of the pitfall trap, Winkler extractor and Berlese funnel for sampling ground-dwelling arthropods in tropical montane cloud forests," Journal of Insect Science, vol. 11, p. 28, 2011.

57. N. V. Meyling and J. Eilenberg, "Isolation and characterisation ofBeauveria bassiana isolates from phylloplanes of hedgerow vegetation,"Mycological Research, vol. 110, no. 2, pp. 188–195, 2006.

58. M. J. Bidochka, A. M. Kamp, T. M. Lavender, J. Dekoning, and J. N. A. De Croos, "Habitat association in two genetic groups of the insect-pathogenic fungus Metarhizium anisopliae: uncovering cryptic species?"Applied and Environmental Microbiology, vol. 67, no. 3, pp. 1335–1342, 2001.

59. M. A. de Muro, S. Mehta, and D. Moore, "The use of amplified fragment length polymorphism for molecular analysis of Beauveria bassianaisolates from Kenya and other countries, and their correlation with host and geographical origin," FEMS Microbiology Letters, vol. 229, no. 2, pp. 249–257, 2003.

60. M. A. de Muro, S. Elliott, D. Moore et al., "Molecular characterisation ofBeauveria bassiana isolates obtained from overwintering sites of Sunn Pests (Eurygaster and Aelia species)," Mycological Research, vol. 109, no. 3, pp. 294–306, 2005.

61. G. D. Inglis, G. M. Duke, M. S. Goettel, and J. T. Kabaluk, "Genetic diversity of Metarhizium anisopliae var. anisopliae in southwestern British Columbia," Journal of Invertebrate Pathology, vol. 98, no. 1, pp. 101–113, 2008.

62. B. H. Bowman, J. W. Taylor, A. G. Brownlee, J. Lee, S. D. Lu, and T. J. White, "Molecular evolution of the fungi: relationship of theBasidiomycetes, Ascomycetes, and Chytridiomycetes," Molecular Biology and Evolution, vol. 9, no. 2, pp. 285–296, 1992.
63. M. Oborník, M. Jirku, and D. Dolezel, "Phylogeny of mitosporic entomopathogenic fungi: is the genus Paecilomyces polyphyletic?"Canadian Journal of Microbiology, vol. 47, no. 9, pp. 813–819, 2001.
64. P. W. Inglis and M. S. Tigano, "Identification and taxonomy of some entomopathogenic Paecilomyces spp. (Ascomycota) isolates using rDNA-ITS sequences," Genetics and Molecular Biology, vol. 29, no. 1, pp. 132–136, 2006.
65. S. A. Rehner and E. P. Buckley, "Isolation and characterization of microsatellite loci from the entomopathogenic fungus Beauveria bassiana (Ascomycota: Hypocreales)," Molecular Ecology Notes, vol. 3, no. 3, pp. 409–411, 2003.
66. J. Enkerli, R. Kölliker, S. Keller, and F. Widmer, "Isolation and characterization of microsatellite markers from the entomopathogenic fungus Metarhizium anisopliae," Molecular Ecology Notes, vol. 5, no. 2, pp. 384–386, 2005.
67. A. E. Arnold and L. C. Lewis, "cology and evolution of fungal endophytes and their roles against insects," in Insect-Fungal Associations: Ecology and Evolution, F. E. Vega and M. Blackwell, Eds., pp. 74–96, Oxford University Press, 2005.
68. G. Hu and J. S. Lager, "Field studies using a recombinant mycoinsecticide (Metarhiziun anisopliae) reveals that it is rhizosphere competent," Applied Environment and Microbiology, vol. 68, pp. 6383–6387, 2002.
69. E. M. Cho, L. Liu, W. Farmerie, and N. O. Keyhani, "EST analysis of cDNA libraries from the entomopathogenic fungus Beauveria (Cordyceps) bassiana. I. Evidence for stage-specific gene expression in aerial conidia, in vitro blastospores and submerged conidia,"Microbiology, vol. 152, no. 9, pp. 2843–2854, 2006.
70. F. M. Freimoser, S. Screen, S. Bagga, G. Hu, and R. J. St Leger, "Expressed sequence tag (EST) analysis of two subspecies ofMetarhizium anisopliae reveals a plethora of secreted proteins with potential activity in insect hosts," Microbiology, vol. 149, no. 1, pp. 239–247, 2003.

71. F. M. Freimoser, G. Hu, and R. J. St Leger, "Variation in gene expression patterns as the insect pathogen Metarhizium anisopliae adapts to different host cuticles or nutrient deprivation in vitro," Microbiology, vol. 151, no. 2, pp. 361–371, 2005.

Chapter 6

Entomopathogenic Nematodes for the Control of Phorid and Sciarid Flies in Mushroom Crops

María Jesús Navarro[1] and Francisco José Gea[1]

[1]Centro de Investigación, Experimentación y Servicios del Champiñón, c/ Peñicas, s/n, 16220, Quintanar del Rey, Cuenca, Spain.

ABSTRACT

The objective of this work was to evaluate the efficacy of two nematodes, *Steinernema feltiae* and *S. carpocapsae*, to control mushroom flies and to evaluate the effect of these treatments on *Agaricus bisporus* production. Two mushroom cultivation trials were carried out in controlled conditions, in substrate previously infested with the diptera

Megaselia halterata and *Lycoriella auripila*, with two treatments: 10^6 infective juveniles (IJ) per square meter of *S. feltiae* and $0.5x10^6$ IJ m^{-2} *S. feltiae* + $0.5x10^6$ IJ m^{-2} *S. carpocapsae*. Another experiment was carried out using the same treatments to evaluate the possible nematode effect on mushroom yield. The number of adults emerging from the substrate was evaluated for each fly species. No decrease in the population of *M. halterata* was detected with nematode application, whereas the number of *L. auripila* was reduced in both treatments, particularly in the individual treatment with *S. feltiae*. The application of entomopathogenic nematodes has no adverse effect on mushroom production.

RESUMO

O objetivo deste trabalho foi avaliar a eficiência de dois nematoides, *Steinernema feltiae* e *S. carpocapsae*, no controle de moscas dos cogumelos e avaliar o efeito desses tratamentos na produção de *Agaricus bisporus*. Foram realizados dois ensaios de cultivo de cogumelos em condições controladas, em substrato previamente infestado pelos dípteros *Megaselia halterata* e *Lycoriella auripila*, com dois tratamentos: 10^6 juvenis infectantes (IJ) por metro quadrado de *S. feltiae* e $0,5x10^6$ IJ m^{-2} de *S. feltiae* + $0,5x10^6$ IJ m^{-2} de *S. carpocapsae*. Foi realizado outro experimento, com uso dos mesmos tratamentos, para avaliar o possível efeito dos nematoides sobre a produção dos cogumelos. Avaliou-se o número de adultos que emergiram no substrato, para cada espécie de mosca. Não foi detectada redução da população de *M. halterata* com a aplicação de nematoides, enquanto o número de *L. auripila*foi reduzido em ambos os tratamentos, particularmente no tratamento individual com *S. feltiae*. A aplicação de nematoides entomopatogênicos não tem efeito adverso na produção de cogumelos.

INTRODUCTION

The phorid Megaselia halterata (Wood) and the sciarid Lycoriella auripila Winnertz are common pests in mushroom farms of Castilla-La Mancha, Spain (Navarro et al., 2002). In their larval stage, these flies feed on mushroom mycelia and even burrow into mushrooms once they are formed (Shamshad, 2010); as adults, they act as vectors of other pests and diseases (Clift et al., 2004; Navarro Lozano et al., 2004; Shamshad et al., 2009; Shamshad, 2010). Unlike in other European countries, where sciarids are present throughout the year and phorids undergo a period of hibernation (Jess et al., 2007), in Spain M. halterata populations are present in high numbers throughout the year, whereas L. auripila is detected in much lower numbers (four times less than M. halterata) and almost exclusively in spring (Navarro et al., 2002).

Traditionally, flies in mushroom crops have been controlled by insecticides, although resistance to certain products has been recorded (Bartlett & Keil, 1997; Smith, 2002). However, the application of phytosanitary products may give rise to two further problems - detrimental effects on the mushroom mycelium, leading to a loss of yield or quality (Grewal et al., 1992; Scheepmaker et al., 1998; Shamshad, 2010), and the presence of residues in the mushrooms once harvested (Navarro & Gea, 2006). Other methods for controlling flies are based on the use of physical barriers to exclude adult individuals from growing in farms (Coles, 2002); on biocontrol organisms, such as mites, bacteria and entomopathogenic nematodes; and, more recently, on plant extracts (Jess & Bingham, 2004;Shamshad et al., 2008; Erler et al., 2009a, 2009b).

Richardson (1987) was the first to use entomo pathogenic nematodes of the genera Steinernema andHeterorhabditis to control mushroom flies. Steinernema spp. nematodes locate and invade fly larvae via anus, mouth or spiracles, and release bacteria associated to them (Xenorhabdus sp. or Photorhabdus sp.), which provokes the death of the infested flies (Kirk & Keil, 2001). However, apart from their compatibility with the insecticides used (Koppenhöfer & Grewal, 2005), the effectiveness of nematodes in mushroom culture depends on other factors such as temperature, moisture and CO2 levels, which may also affect them (Kirk & Keil, 2001).

Information about phorid control by nematodes is scarce in the literature. According to Grewal et al. (1993), the effect of S. feltiae on M. halterata is not significant. However, Scheepmaker et al. (1997) found a substantial decrease in the second generation of M. halterata following the application of 3x106 infective juveniles (IJ) per square meter of S. feltiae one week after casing. More recently, Erler et al. (2009b) described the effectiveness ofS. feltiae applied on substrates after controlled infestation with M. halterata. However, S. carpocapsae, applied at a dose of up to 15x106 IJ m-2 on casing soil, coinciding with the third larval instar, can reduce phorid populations by up to 73% (Jess & Bingham, 2004). This concentration, however, is too expensive to have any commercial appeal. For the control of sciarids with nematodes, many authors defend the use of S. feltiae, having observed effectiveness between 66 and 95%, depending on the fly population and the timing of nematode application (Scheepmaker et al., 1997; Shamshad et al., 2008). In the case of S. carpocapsae, Gouge & Hague (1995) reported the relative inefficacy of this species against sciarids, whereas other authors defend its efficacy, but with a lower infectivity than that of S. feltiae (Kim et al., 2004). Furthermore, some authors observed a detrimental effect of treatments involving entomophathogenic nematodes on mushroom mycelia, such as reduced yield in early flushes, depending on the nematode dosage rate (Grewal et al., 1992).

The objective of this work was to evaluate the efficacy of two nematodes, *S. feltiae* and *S. carpocapsae*, for control of mushroom flies and to evaluate the effect of these treatments on *Agaricus bisporus* production.

MATERIALS AND METHODS

Two cropping trials were set up in an experimental mushroom growing room (20 m3 approximately) following the standard practices used in mushroom farms in Spain (Navarro Lozano et al., 2004). In each trial, A. bisporus(Lange) Imbach, Gurelan 45 strain, (Gurelan S. Coop., Huarte, Pamplona, Spain) was cultivated in 40 experimental trays (16 L in volume, 870 cm2 in area) filled with 6 kg of commercial mushroom compost (Villacasa S.L., Casasimarro, Cuenca, Spain) spawned at 0.9%. Spawn run took place for 15 days at 26°C and 95% relative humidity (RH). On day 15, trays were cased with a 40 mm layer of Topterra mushroom casing (Topterra Holland, Wanssum, Netherlands), 3.5 L per tray, previously hydrated to 72-76% moisture. The environmental conditions were modified to stimulate the production of mushroom fruit bodies. A temperature of 18°C and RH of 85-90% were maintained throughout harvest. The trials were concluded after 50 days. A randomized complete block design, with four treatments and ten replicates, was used in each trial. The treatments applied were: Sf, *S. feltiae* treatment, fly-infested trays and subsequent application of 10^6 IJ m^{-2} *S. feltiae*, Biorend R champiñones, (Idebio S.L., Salamanca, Spain) together with 1 mL of chitosan in 150 mL water per tray; Sf+Sc, *S. feltiae* + *S. carpocapsae* treatment, infested trays and subsequent application of 0.5×10^6 IJ m^{-2} *S. feltiae* + 0.5×10^6 IJ m^{-2} *S. carpocapsae*, Biorend R champiñones + Biorend R palmeras, (Idebio S.L., Salamanca, Spain) together with 1 mL of chitosan in 150 mL water per tray. Ten trays free of fly infestation and with no nematode application were used as non-infested controls (C), and ten fly-infested trays with no nematode application were used as infested controls (IC).

The first trial was performed to evaluate the effect of the treatments on fly control under an infection pressure of 800 phorids and 52 sciarids per day. On day 5 of spawn running, the trays corresponding to treatments IC, Sf, and Sf+Sc were exposed to flies for 48 hours in a commercial mushroom growing house of 45 m^2 cropping area. Then, all the trays were covered with a cubic structure (37x37x37 cm) made of anti-trip (0.6 mm) mesh and placed back in the experimental mushroom growing room to continue spawn running. A yellow sticky trap (14x7.5 cm) was placed inside each of these cubic structures to capture the flies. On day 16, the different entomopathogenic nematode treatments were applied on the surface of each tray. In this way, the time elapsing between natural infestation and the application of nematodes was ten days.

The second trial was performed to evaluate the effect of the treatments on fly control under an infection pressure of 1,237 phorids and 82 sciarids per day. On day 11 of spawn running, the trays corresponding to treatments IC, Sf, and Sf+Sc were exposed to infestation for 48 hours in a commercial mushroom growing house of 45 m^2 cropping area. Subsequently, all the trays were handled as in the first trial. On day 16, the different entomopathogenic nematode treatments were applied on the surface of each tray. In this case, the time elapsing between natural infestation and the application of nematodes was four days.

To prevent the flies from escaping, harvesting of the first flush was delayed until the sticky traps had been withdrawn. The experiments were considered finished after the emergence of all the first generation adults. Then, the captured flies were counted and identified by binocular stereomicroscope. Treatments were evaluated by counting the emerged adults per tray in each treatment. The percentage reduction in adult emergence for each treatment in each of the trials was calculated by the formula: % Reduction = [(Y - X)/Y] x 100, in which X is the number of emerging adults in each of the treatments, and Y is the number of emerging adults in the infested control (Erler et al., 2009a).

To determine the effect of the entomopathogenic nematodes on *A. bisporus* yield, a cropping trial was set up in an experimental mushroom growing room, under the same conditions as the two previous trials but with no dipteral infestations. Two treatments and a control were also established: Sf, Sf+Sc, and C. Mushrooms were harvested daily for two flushes. The number and total weight of the fruit bodies were recorded for each treatment. The effect of entomopathogenic nematodes on mushroom productivity was evaluated from the total yield of harvested mushrooms, and from the number and unitary weight of mushrooms. The effect of treatments on mushroom productivity was also evaluated from the biological efficiency of the crop, calculated as the ratio of the fresh weight of the total yield of harvested mushrooms to the weight of dry substrate at spawning, expressing the fraction as kg $100kg^{-1}$ compost. In addition, the earliness of each treatment was expressed as the number of days between casing and harvesting of the first flush. This experiment was set up as a randomized complete block design with 12 replicates.

Analyses of variance, with logarithmic or square root transformation for count data and angular transformation of the percentage data to stabilize variances when necessary, were done using the software package Statgraphics Plus, version 4.1 (Statistical Graphics Corp., Princeton, NJ, USA). Tukey's test, at 5% probability, was used to establish significant differences between means.

RESULTS AND DISCUSSION

The total number of flies captured in the first trial was lower than in the second one, reflecting the lower degree of initial infection pressure in the first trial. Although fly captures were registered in the non-infested trays (Table 1), in both trials and for both flies, the data obtained for treatment IC were significantly higher than those for C, indicating the good progress of the infection process.

In the first trial, the mean number of emerging *M. halterata* was 233 for the infested control trays (IC), and 186.3 and 221.8 for the nematode treatments Sf and Sf+Sc, respectively, with no significant differences between them ($F_{2,29} = 0.63$; $p = 0.5415$). No significant differences ($F_{2,23} = 1.21$; $p = 0.3168$) were observed between the numbers of phorid flies captured in the treatments IC, Sf, and Sf+Sc in the second trial (Table 1). There was also no effect of the treatments Sf and Sf+Sc in the emergence of *M. halterata* adults in both trials, and no phorid percentage reduction was verified for any treatment.

Table 1: Mean numbers±standard error of adult Megaselia halterata and Lycoriella auripila emerging from the different trays, and the percentage reduction of L. auripila obtained for each treatment in both trials (1)

Treatment	M. halterata		L. auripila		L. auripila reduction (%)(2)	
	First trial	Second trial	First trial	Second trial	First trial	Second trial
C	14.1±6.9a	22.3±18.0a	2.0±2.2a	40.7±13.2a	-	-
IC	233.0±97.6b	462.7±166.2b	136.2±60.7c	184.4±46.7c	-	-
Sf	186.3±57.3b	366.1±80.3b	50.3±26.2b	116.2±35.4b	63.1B	37.0A
Sf+Sc	221.8±121.6b	406.3±133.2b	68.1±47.4b	180.0±53.2c	50.0B	2.4A

[1]For each trial, means followed by equal letters, lowercase in the columns, do not differ significantly by Tukey's test, at 5% probability

[2]Means followed by equal letters, uppercase in the line, do not differ significantly by Tukey's test, at 5% probability. C, non infested control; IC, infested control without nematode treatments; Sf, nematode application of 10^6 IJ m^{-2} *Steinernema feltiae*; Sf+Sc, nematode application of $0.5x10^6$ IJ m^{-2} *S. feltiae* + $0.5x10^6$ IJ m^{-2} *S. carpocapsa*

The ineffectiveness of *S. feltiae* for controlling phorids agrees with most of the consulted bibliography (Grewal et al., 1993; Jess et al., 2005). However, it contrasts with the findings of Long et al. (2000) in laboratory bioassays. These authors reported that *S. feltiae* is effective in the control of *M. halterata*, probably because nematodes applied on growing substrates show different behavior. The results of the present work also contrast with the findings of other authors, who observed a reduction of more than 70% in the emergence of *M. halterata* adults after the application of *S. feltiae* (Scheepmaker et al., 1997; Erler et al., 2009b). This divergence in results was probably due to the fact that in the present study only the first generation of adults after infestation was considered, whereas the above authors also considered subsequent generations.

The ineffectiveness of the Sf+Sc treatment on *M. halterata* was observed regardless of the trials, which contrasts with the results obtained by Jess & Bingham (2004), who confirmed the use of *S. carpocapsae* to control phorids. This discrepancy might be due to the lower application doses used or to the application of both nematode species.

In trial conditions, the inefficacy of both treatments against *M. halterata* was independent of the time elapsing between infestation and nematode application. For optimum efficacy, nematodes should be applied when a majority of hosts are susceptible to nematode infection. In the first trial, considering the environmental conditions and the time elapsing between infestation and treatment (9-10 days), the larvae were presumed to be in a sufficiently advanced stage (third larval instar) to allow infestation by nematodes (Jess & Bingham, 2004; Jess et al., 2005). However, not even in these favorable conditions did the applications lead to an acceptable level of reduction when the nematode formulations were used. Further efforts to develop entomopathogenic nematodes against *M. halterata* are necessary.

In the first trial, the mean number of emerging *L. auripila* was 136 for the infested control trays (IC), and significantly lower ($F_{2,29} = 9.42$; $p = 0.0007$) for both nematode treatments. There was no significant difference between the Sf and Sf+Sc treatments (Table 1). In the second trial, 184 sciarid flies were captured emerging from the infested control treatment, a value statistically similar to that recorded for the Sf+Sc treatment; however, this number was significantly reduced ($F_{2,23} = 6.53$; $p = 0.0057$) in the Sf treatment (116 sciarid flies). In this trial, only the Sf treatment reduced the adult emergence of *L. auripila* when compared to the infested control. Gouge & Hague (1995) established that *S. carpocapsae* was relatively ineffective against different sciarid species. The results obtained in the present work are in agreement with this, but contradict those of Kim et al. (2004), who found that *S. carpocapsae* can be an effective tool for the management of the fungus gnat larvae. Both nematode species were applied together so that each could combat its own target pest. However, there was no improvement in the efficacy of the combined treatment against either fly species targeted.

For the control of sciarids (Table 1), the reduction percentage (63%) obtained with the Sf treatment in the first trial can be considered satisfactory; however, it was less than the 80-95% reduction in sciarids obtained by other authors (Scheepmaker et al., 1997; Shamshad et al., 2008; Jess & Schweizer, 2009). In the second trial, the reduction percentage (37%) was significantly lower than that in the first trial ($F_{1,18}$ = 9.15; p = 0.0073). In the case of the Sf+Sc treatment, the reduction percentage obtained in first trial (50%) was also satisfactory and significantly higher than the one in the second trial ($F_{1,13}$ = 19.41; p = 0.0007). Efficacy was lower for each treatment in the second trial. Considering the environmental conditions and the time elapsing between infestation and treatment (3-4 days), larvae were probably not large enough to be parasitized. This agrees with the recommendation of Scheepmaker et al. (1997) to postpone treatment to a week after infestation. A longer time between infestation and nematode application would favor fly control because the nematodes would find larger larvae (third and fourth stage), which are more vulnerable to *S. feltiae* attack (Jess & Bingham, 2004; Jess et al., 2005)

The mean values of the mushroom yields, per flush and total yield, for each of the treatments (C, Sf, and Sf+Sc) were statistically similar for the first flush ($F_{2,31}$ = 1.60; p = 0.2188), the second flush ($F_{2,31}$ = 2.98; p = 0.0657), and for total harvest ($F_{2,31}$ = 0.31; p = 0.7346); a total yield of around 20 kg m^{-2} was obtained in all three cases (Figure 1). This result reflects the statistically similar biological efficiency ($F_{2,31}$ = 0.31; p = 0.7343) of all the treatments, with values close to 85 kg of mushrooms per 100 kg of compost (dry weight) collected over two flushes (Table 2).

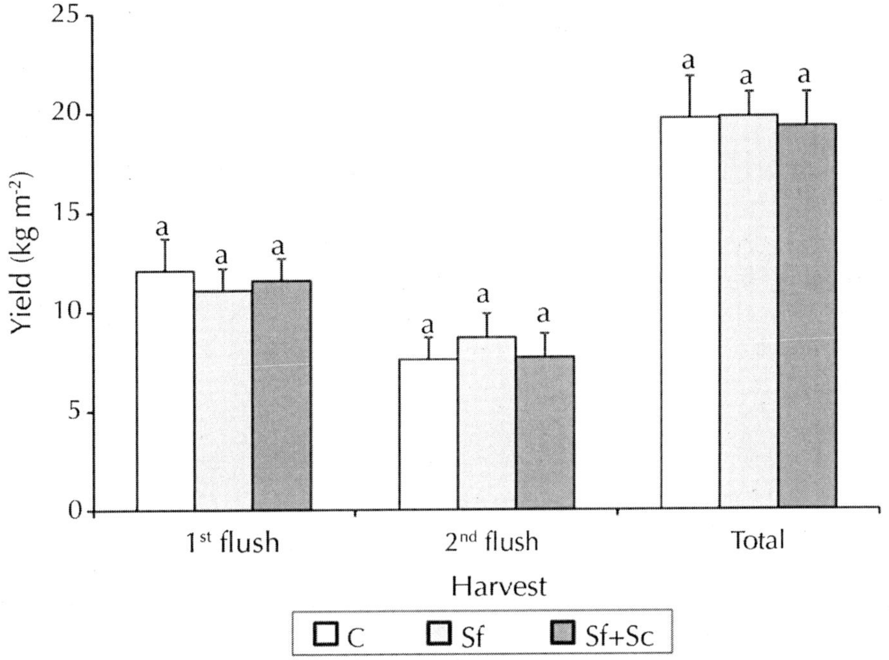

Figure 1: Mushroom yield, expressed per flush and total yield, for each of the treatments evaluated. C, control; Sf, trays with nematode application of 106 IJ m-2 Steinernema feltiae; Sf+Sc, trays with nematode application of 0.5x106 IJ m-2 S. feltiae + 0.5x106 IJ m-2 S. carpocapsae. For each harvest, means followed by equal letters do not differ significantly by Tukey's test, at 5% probability.

Table 2: Biological efficiency (kg of mushroom per 100 kg of dried compost), unitary weight, and earliness (days to first flush harvest) (1)

Treatments	Biological efficiency	Unitary weight	Earliness
C	84.8a	12.5a	21.5a
Sf	85.3a	15.4b	21.6a
Sf+Sc	83.0a	12.7a	21.7a

[1]Means followed by equal letters, in the columns, do not differ significantly by Tukey's test, at 5% probability. C, non-infested control; Sf, nematode application of

10^6 IJ m^{-2} *Steinernema feltiae*; Sf+Sc, nematode application of 0.5×10^6 IJ m^{-2} *S. feltiae* + 0.5×10^6 IJ m^{-2} *S. carpocapsae*.

Regarding the number of mushrooms harvested, both treatments showed lower numbers than the control in the first flush. In the Sf+Sc treatment this decrease was compensated in the second flush since the total results for total yield only showed a significant drop ($F_{2,31}$ = 5.28; p = 0.0107) for the Sf treatment (Figure 2). Moreover, the unitary weight of the mushrooms was significantly higher ($F_{2,31}$ = 10.27; p = 0.0001) for the Sf treatment than for the control and for Sf+Sc (Table 2). Lastly, earliness, defined as the time between applying the casing mixture and the first flush, varied between 21.5 and 21.7 days, with no significant differences ($F_{2,31}$ = 0.28; p = 0.7577) between treatments.

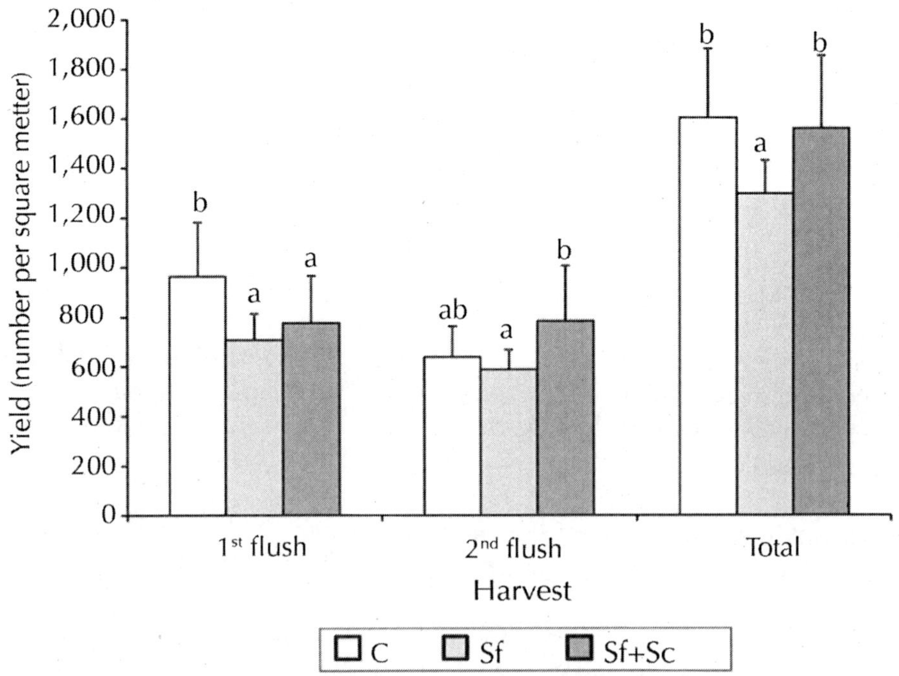

Figure 2: Mushroom yield, expressed per flush and total yield, for each of the treatments evaluated. C, control; Sf, trays with nematode application of 106 IJ m-2 Steinernema feltiae; Sf+Sc, trays with nematode application of 0.5 x 106 IJ m-2 S. feltiae + 0.5 x 106 IJ m-2 S. carpocapsae. For each harvest, means followed by equal letters do not differ significantly by Tukey's test, at 5% probability.

Some authors maintain that the use of entomopathogenic nematodes affects mycelial growth, depending on the nematode dosage rate (Grewal et al., 1992), leading to decreased yields for the first flush, although subsequent flushes typically compensate for this early yield loss (Jess et al., 2005). The results found in the present work contradict this affirmation since the overall yields were similar in both flushes for all treatments. For the number of carpophores collected, Grewal et al. (1992) obtained 20% more mushrooms after the application of *S. feltiae*, which was attributed to the increased dispersion of the bacterium *Pseudomonas putida*, responsible for the formation of primordia in the casing material. In the present study, the application of *S. feltiae* alone reduced the number of mushrooms harvested, but the weight of individual sporophores was higher, which might be considered advantageous for better maintaining post-harvest quality, especially in the first flush when there are normally too many mushrooms to pick (Scheepmaker et al., 1998).

CONCLUSIONS

The application of entomopathogenic nematodes has no adverse effect on mushroom production.

The application of *Steinernema feltiae* ten days after the foreseen time of infestation is beneficial for the control of sciarids.

The application of *S. feltiae*, alone or with *S. carpocapsae*, has no effect on the phorids found in mushroom crops.

REFERENCES

1. BARTLETT, G.R.; KEIL, C.B.O. Identification and characterization of a permethrin resistance mechanism in populations of the fungus gnat Lycoriella mali (Fitch) (Diptera: Sciaridae). Pesticide Biochemistry and Physiology, v.58, p.173-181, 1997. DOI: 10.1006/pest.1997.2306.
2. CLIFT, A.; SHAMSHAD, A.; TERRAS, M.A. Flies and dry bubble on cultivated mushrooms. In: ROMAINE, C.P.; KEIL, C.B.; RINKER, D.L.; ROYSE, D.J. (Ed.). Science and cultivation of the edible and medicinal fungi. State College: The Pennsylvania State University, 2004. p.459-464.

3. COLES, P. Integrated pest management in mushroom production: specific control techniques: exclusion. In: MUSHROOM integrated pest management. State College: The Pennsylvania State University, 2002. p.21-26.
4. ERLER, F.; POLAT, E.; DEMIR, H.; CETIN, H.; ERDEMIR, T. Control of the mushroom phorid fly, Megaselia halterata (Wood), with plant extracts. Pest Management Science, v.65, p.144-149, 2009a. DOI: 10.1002/ps.1658.
5. ERLER, F.; POLAT, E.; DEMIR, H.; CETIN, H.; ERDEMIR, T. Evaluation of microbial products for the control of the mushroom phorid fly, Megaselia halterata (Wood). Journal of Entomological Science, v.44, p.89-97, 2009b.
6. GOUGE, D.H.; HAGUE, N.G.M. The susceptibility of different species of sciarid flies to entomopathogenic nematodes. Journal of Helminthology, v.69, p.313-318, 1995. DOI: 10.1017/S0022149X00014887.
7. GREWAL, P.S.; RICHARDSON, P.N.; COLLINS, G.; EDMONDSON, R.N. Comparative effects of Steinernema feltiae (Nematoda: Steinernematidae) and insecticides on yield and cropping of the mushroom Agaricus bisporus. Annals of Applied Biology, v.121, p.511-520, 1992. DOI: 10.1111/j.1744-7348.1992.tb03461.x.
8. GREWAL, P.S.; TOMALAK, M.; KEIL, C.B.O.; GAUGLER, R. Evaluation of a genetically selected strain of Steinernema feltiae against the mushroom sciarid Lycoriella mali. Annals of Applied Biology, v.123, p.695-702, 1993. DOI: 10.1111/j.1744-7348.1993.tb04939.x.
9. JESS, S.; BINGHAM, J.F.W. Biological control of sciarid and phorid pests of mushroom with predatory mites from the genus Hypoaspis (Acari: Hypoaspidae) and the entomopathogenic nematode Steinernema feltiae. Bulletin of Entomological Research, v.94, p.159-167, 2004. DOI: 10.1079/BER2003286.
10. JESS, S.; MURCHIE, A.K.; BINGHAM, J.F.W. Potential sources of sciarid and phorid infestations and implications for centralised phases I and II mushroom compost production. Crop Protection, v.26, p.455-464, 2007. DOI: 10.1016/j.cropro.2006.04.015.
11. JESS, S.; SCHWEIZER, H. Biological control of Lycoriella ingenua (Diptera: Sciaridae) in commercial mushroom (Agaricus

bisporus) cultivation: a comparison between Hypoaspis miles and Steinernema feltiae. Pest Management Science, v.65, p.1195-1200, 2009. DOI: 10.1002/ps.1809.

12. JESS, S.; SCHWEIZER, H.; KILPATRICK, M. Mushroom applications. In: GREWAL, P.S.; EHLERS, R.-U.; SHAPIRO-ILAN, D.I. (Ed.). Nematodes as biocontrol agents. New York: CABI Publishing, 2005. p.191-213. DOI: 10.1079/9780851990170.0191.

13. KIM, H.H.; CHOO, H.Y.; KAYA, H.K.; LEE, D.W.; LEE, S.M.; JEON, H.Y. Steinernema carpocapsae (Rhabditida: Steinernematidae) as a biological control agent against the fungus gnat Bradysia agrestis (Diptera: Sciaridae) in propagation houses. Biocontrol Science and Technology, v.14, p.171-183, 2004. DOI: 10.1080/09583150310001655693.

14. KIRK, D.J.; KEIL, C.B. Factors influencing efficacy of two entomopathogenic nematodes used for fly control in commercial mushroom crops. Mushroom News, v.49, p.4-17, 2001.

15. KOPPENHÖFER, A.M.; GREWAL, P.S. Compatibility and interactions with agrochemicals and other biocontrol agents. In: GREWAL, P.S.; EHLERS, R.-U.; SHAPIRO-ILAN, D.I. (Ed.). Nematodes as biocontrol agents. New York: CABI Publishing, 2005. p.363-381. DOI: 10.1079/9780851990170.0363.

16. LONG, S.J.; RICHARDSON, P.N.; WILLMOTT, D.M.; EDMONTON, R.N. Infectivity of entomopathogenic nematodes (Steinernematidae, Heterorhabditidae) to mushroom phorid fly (Megaselia halterata) larvae. Nematology, v.2, p.451-459, 2000. DOI: 10.1163/156854100509312.

17. NAVARRO, M.J.; ESCUDERO, A.; FERRAGUT, F.; GEA, F.J. Evolution and seasonal abundance of phorid and sciarid filies in Spanish mushroom crops. In: SÁNCHEZ, J.E.; HUERTA, G.; MONTIEL, E. (Ed.). Mushroom biology and mushroom products. Jalisco: Universidad Autónoma del Estado de México, 2002. p.189-195.

18. NAVARRO, M.J.; GEA, F.J. Estudio de la fitotoxicidad del insecticida diflubenzuron en el cultivo de champiñón. Determinación del nivel de residuos. Boletín de la Asociación Española de Cultivadores de Champiñón, v.48, p.32-34, 2006.

19. NAVARRO LOZANO, M.J.; GEA ALEGRÍA, F.J.; FERRAGUT PÉREZ, F. Biología y control del ácaro miceliófago Brennandania

lambi (Krczal) en los cultivos de champiñón de Castilla-La Mancha. Madrid: Ministerio de Agricultura, Pesca y Alimentación, 2004. 203p.

20. RICHARDSON, P.N. Susceptibility of mushroom pests to the insect-parasitic nematode Steinernema feltiae and Heterorhabditis heliothidis. Annals of Applied Biology, v.111, p.433-438, 1987. DOI: 10.1111/j.1744-7348.1987.tb01471.x.

21. SCHEEPMAKER, J.W.A.; GEELS, F.P.; RUTJENS, A.J.; SMITS, P.H.; VAN GRIENSVEN, L.J.L.D. Influence of Steinernema feltiae and diflubenzuron on yield and economics of the cultivated mushroom Agaricus bisporus in Dutch mushroom culture. Biocontrol Science and Technology, v.8, p.269-275, 1998. DOI: 10.1080/09583159830333.

22. SCHEEPMAKER, J.W.A.; GEELS, F.P.; SMITS, P.H.; VAN GRIENSVEN, L.J.L.D. Control of the mushroom pests Lycoriella auripila (Diptera: Sciaridae) and Megaselia halterata (Diptera: Phoridae) by Steinernema feltiae (Nematoda: Steinernematidae) in field experiments. Annals of Applied Biology, v.131, p.359-368, 1997. DOI: 10.1111/j.1744-7348.1997.tb05165.x.

23. SHAMSHAD, A. The development of integrated pest management for the control of mushroom sciarid flies, Lycoriella ingenua (Dufour) and Bradysia ocellaris (Comstock), in cultivated mushrooms. Pest Management Science, v.66, p.1063-1074, 2010. DOI: 10.1002/ps.1987.

24. SHAMSHAD, A.; CLIFT, A.D.; MANSFIELD, S. Toxicity of six commercially formulated insecticides and biopesticides to third instar larvae of mushroom sciarid, Lycoriella ingenua Dufour (Diptera: Sciaridae), in New South Wales, Australia. Australian Journal of Entomology, v.47, p.256-260, 2008. DOI: 10.1111/j.1440-6055.2008.00653.x.

25. SHAMSHAD, A.; CLIFT, A.D.; MANSFIELD, S. The effect of tibia morphology on vector competency of mushroom sciarid flies. Journal of Applied Entomology, v.133, p.484-490, 2009. DOI: 10.1111/j.1439-0418.2008.01362.x.

26. SMITH, J.E. Dimilin resistance in mushroom sciarids. The Mushroom Journal, v.626, p.15, 2002.

Chapter 7

Pathogens Associated with Sugarcane Borers, Diatraea spp. (Lepidoptera: Crambidae): A Review

Víctor M. Hernández-Velázquez[1], Laura P. Lina-García1, Verónica Obregón-Barboza[1], Adriana G. Trejo-Loyo[2], and Guadalupe Peña-Chora[2]

[1]Centro de Investigación en Biotecnología, Universidad Autónoma del Estado de Morelos, Avenida Universidad No. 1001, Colonia Chamilpa, 62210 Cuernavaca, MOR, Mexico

[2]Centro de Investigaciones Biológicas, Universidad Autónoma del Estado de Morelos, Avenida Universidad No. 1001, Colonia Chamilpa, 62210 Cuernavaca, MOR, Mexico

ABSTRACT

The objective of this paper was to analyze information related to entomopathogenic-associated Diatraea spp. Gaining a better understanding of the effects of these microorganisms will help in the development of successful microbial control strategies against stem borers that attack sugarcane plants.

INTRODUCTION

The Diatraea spp. (Lepidoptera: Crambidae) complex is only found in the American continent, and it is the most important group of stem borers that principally attack maize and sugarcane, as well as other gramineous crops, including rice, sorghum, and forage grasses [1]. The sugarcane borer (SCB) D. saccharalis Fab. is the most economically important pest in South America [2, 3], whereas the neotropical corn stalk borer (NCB) D. lineolata Walk is primarily found in Central America [4], D. magnifactella Dyar and D. considerata Heinrich are found in Mexico [1], and the southwestern corn borer D. grandiosella Dyar is found in the United States [5].

Unfortunately, commercially available insecticides are not efficient for the control of Diatraea spp. for a variety of reasons, mainly because of the continuous presence of the host plants in fields throughout the year, the concomitant presence of mature and immature forms of the insect, and the cryptic feeding habits of the insect [6]. An alternative strategy is integrated pest management with biological control as the first defense, which includes the use of parasitoids and entomopathogens.

In this paper, we present a perspective on the attempts to control Diatraea spp. using pathogens in the Americas. We also discuss the status of recent attempts to use pathogens in the field.

FUNGI

One promising field for research is the use of entomopathogenic fungi as biological control agents of insect pests in sugarcane plants. Approximately 80% of the etiological agents involved in insect diseases are fungi, which encompass 90 genera and more than 700 species [7]. A number of fungi (Hypocreales: Clavicipitaceae) including B. bassiana, B. brongniartii, M. anisopliae, P. fumosoroseus, Hirsutella sp., Cylindrocarpon sp., andNomuraea rileyi have been isolated from Diatraea spp. in the Americas from Argentina to the USA (Table 1). Under certain climatic conditions, B. bassiana has been reported to cause natural epizootics on D. grandiosella[8].

Table 1: Entomopathogens fungi from Diatraea spp

Host species	Entomopathogen	Country	Reference
D. grandiosella D. crambidoides	B. bassiana (1–8%)	Mississippi, United States	Inglis et al., 2000 [5]
D. grandiosella	B. bassiana (epizootics)	Texas High Plains, United States	Knutson and Gilstrap, 1990 [8]
	B. bassiana		
D. saccharalis	M. anisopliae	Venezuela	Zambrano et al., 2002 [9]
	P. fumosoroseus		
D. saccharalis	M. anisopliae	PE, Brazil	Alves, et al. 2002 [10]
D. saccharalis	B. bassiana	Pinar del Río, La Habana, Matanzas, Villa Clara, Cienfuegos and Camagüey, Cuba	Estrada et al., 2004 [11]
D. saccharalis	B. bassiana M. anisopliae Nomuraea rileyi Iseria sp.	Tucumán, Argentina	Yasen de Romero et al., 2008 [12]
D. saccharalis	M. anisopliae (Ma2 and Ma3)	Mexico	Angel-Sahagún et al., 2005 [13]
	B. bassiana		
	ARSEF: 1489	Pernambuco	
	1832	Brazil	
	1834	Brazil	
	2629	Brazil	
	3020	Pernambuco	
	3858	Brazil	

D. saccharalis	5500	Pernambuco,	Humber et al., 2009 [14]
	5502	Brazil	
		Brazil	
		Santa Fe,	
		Argentina	
		Santa Fe,	
		Argentina	
D. saccharalis	B. brigniartii ARSEF: 1830	Sao Paulo, Brazil	Humber et al., 2009 [14]
D. saccharalis	Cylindrocarpon sp ARSEF: 8043 8044	Colima, Mexico	Humber et al., 2009 [14]
	M. anisopliae	Colima, Mexico	
	ARSEF: 3290	Colima, Mexico	
D. saccharalis	3291	Colima, Mexico	Humber et al., 2009 [14]
	3292	Colima, Mexico	
	3298	Colima, Mexico	
	M. anisolpliae		
D. magnifactella	B. bassiana	Mexico	Hernández and Velázquez 2004 [15]
	Hirsutella sp.		
D.grandiosella	Nosema sp. isolate 167	Marshal County, (Mississippi, USA)	Inglis et al. 2000 [5]
D.grandiosella	Nosema sp. isolates 295, 504	Oktibbeha County, (Mississippi, USA)	Inglis et al. 2000 [5]
D.grandiosella	Nosema sp. isolates 181, 513, 522	Washington County, (Mississippi, USA)	Inglis et al. 2000 [5]

D. saccharalis	Nosema sp.	Colombia	Lastra and Gómez 2000 [16]
D. saccharalis	Granulovirus (DsGV)	Southern United States	Pavan et al. (1983) [17]
D. saccharalis	Densovirus (DsDNV)	Guadeloupe	Meynadier et al. (1977) [18]
D. magnifactella	B. thuringiensis	Mexico	Fonseca-González et al. (2011) [19]

The life cycle of entomopathogenic fungi in the arthropod hosts is initiated with the germination of conidia that contacts the host integument and produces a germ tube that penetrates the host through a combination of physical pressure and enzymatic degradation of the cuticle. The fungus initially colonizes the host through a yeast phase. Host death usually results from a combination of nutrient depletion, invasion of organs, and the action of fungal toxins. Hyphae usually emerge from the cadaver. The mummified corpse of the insect remains in the environment for several weeks and, in the case of stem borer, it keeps remains protected inside the stem. Therefore, it is more common to detect and isolate fungi compared to other pathogens that destroy the host, such as Bacillus thuringiensis (Bt), viruses, or nematodes.

Entomopathogenic fungi are widely distributed in all regions of the world; these species have wide genetic variation among the different isolates. Pathogenicity and virulence to different species, as well as enzymatic and DNA characteristics, vary among different isolates [14, 20, 21]. Therefore, it is important to evaluate as many geographic isolates as possible from different areas to select the most suitable isolate based on its virulence and growth at high temperatures. Several research groups have verified the pathogenicity and virulence of Hypocreales fungi, such as M. anisopliae and B. bassiana (Table 2), which have become important biocontrol agents used for the microbial control of Diatraea spp. It is possible that a limited selection of available isolates of B. bassiana could identify highly virulent strains from each of the different Diatraeaspecies.

Table 2: Entomopathogenic-fungi bioassays on Diatraea spp

Host species	Pathogen	Bioassay	Result	Reference
D. saccharalis	B. bassiana	Immersing third instar larvae into a suspension of 108 conidia/ mL	TL50 (conidia isolates Bb1 and 5): 4.3 days. Mortality dry mycelium preparations: 21.3% (Bb1) 82.5% (Bb5) at 7 days after inoculation	Arcas et al., 1999 [26]
D. saccharalis	B. bassiana Mycotrol strain GHA	Sprayed first, second and third instar larvae with 1 mL	LD50 in spores/mm2 1st instar: 72.1 2nd instar: 384.3 3rd instar: 777.0 Mean days of survival 1st instar: 4.6 2nd instar: 4.8 3rd instar: 6.4	Legaspi et al., 2000 [27]
D. grandiosella	B. bassiana Eight isolates	Dipped in conidial suspension (dosage not specified)	Mortality from 10 to 21%	Inglis et al., 2000 [5]

D. saccharalis	B. bassiana ATCC 20872 from Solenopsis invicta	Third instar larvae sprayed with suspension of 108 conida/ mL	TL50 3.02 to 4.10 days	Marques et al., 2000 [28]
D. sacccharalis	B. bassiana ATCC 20872 from Solenopsis invicta	Third instar larvae sprayed with 3 mL of yeast-like cells or conidia	LC50 5.6 × 106 yeast/mL and 4.8 × 106 conidia/ mL	Alves et al., 2002 [29]
D. saccharalis	B. bassiana IBCB from Brazil	1 cm larvae sprayed with 1 mL	CL50 1.58 × 107 conidia/ mL at 6 days after inoculation	Wenzel et al., 2006 [30]

Field evaluations have been performed using M. anisopliae and B. bassiana against several insect pests of sugarcane, including D. saccharalis in Brazil. Application of M. anisopliae at a rate of 1 × 10^{13} spores per hectare caused 58% mortality of D. saccharalis, and B. bassiana at 3.7 × 10^{8} spores per milliliter reduced D. saccharalis damage by 45% (Alves et al., 1984 and 1985, cited in Legaspi et al. [22]). For effectiveness in the field, it is important to consider the contact between the spores and the host, formulation, and the virulence of the pathogens.

MICROSPORIDIA

Microsporidia (Eukaryota: Fungi) is the most ubiquitous group among insect populations [23, 24]. Microsporidia are tiny unicellular organisms (from 2 to 40 μm in diameter), that are opportunistic and obligate intracellular parasites and attack different groups of invertebrate and vertebrate animals. Microsporidia generally produce chronic diseases and reduce the physiological and reproductive ability of their hosts. Many species of microsporidia infect arthropods, especially insects such as Lepidoptera and Coleoptera [23–25].

In general, these microorganisms live as parasites in cells of the midgut epithelium, where they complete their development, and the cycle starts when the infective states of microsporidia (spores) arrive at the digestive tube and colonize this region all the way to the excretory system. The spores germinate because of the acid intestinal pH, the microorganisms penetrate the midgut cells, and the intestinal activity is paralyzed between 14 to 21 days later because the insect cannot assimilate nutrients. Nosema locustae infects the adipocytes of fat body, which interferes with the adequate function of the insect's intermediary metabolism and competes with the insect for energetic reserves. Microsporidia produce effects that depend on the species and concentration; however, they generally produce weakness and eventually lead to death [31, 32]. Only some species in this group have the possibility to be potentially relevant for natural or classical control. There are many studies worldwide on the pathogenic effects of the microsporidia N. pyrausta (Payllot) andVairimorpha necatrix (Kramer) on borer organisms such as Ostrinia nubilalis (Hübner) (European corn borer), Lymantria dispar (Gypsy moth), and the grasshopper [5, 32]. In this group, only the microsporidiumN. locustae has been registered as a microbial insecticide for the control of grasshoppers in grasslands [33]. In many areas of the USA and Europe, Nosema is the main agent used for the control of grasshoppers; however, in Latin America, few studies have been performed with Nosema [33]. For example, Inglis et al. [5] reported six isolates of Nosema on larvae in the winter diapause stage from D. grandiosella Dyer collected during 1998 in corn stems from three locations in Mississippi, USA. However, the frequency of infection in the field was very low (1, 3, and 15% in the counties of Marshal, Oktibbeha, and Washington, resp.), and no isolates were found in D. crambidoides (Grote) (Table 3).

Table 3: Entomopathogenic-microsporidia bioassays on Diatraea spp

Host specie	Pathogen	Bioassay	Results	Reference
D.grandiosella	Nosema sp isolate 167	Surface of diet (squares), concentration not mentioned. Larvae and pupae	Mortality of 1–15%	Inglis et al., 2000 [5]

D.grandiosella	Nosema sp isolate 295, 504	Surface of diet (squares), concentration not mentioned. Larvae and pupae	Mortality of 1–15%	Inglis et al., 2000 [5]
D.grandiosella	Nosema sp isolate 181, 513, 522	Surface of diet (squares), concentration not mentioned. Larvae and pupae	Mortality of 1–15%	Inglis et al., 2000 [5]
D. saccharalis	Vairimorpha necatrix	1st instar larvae, Surface of diet	Microsporidiosis LC50 = 48.4 spores/mm2 Gut damage LC50 = 8941 spores/ mm2	Fuxa, 1981 [32]
D.grandiosella	Nosema sp isolate 506	5-day-old larvae and pupae, leaf pieces ~5 cm2 with 10l inoculum (101 to 107 spores)	Median infective dose of 2 × 103 spores per larva, pupae small, decreased egg production	Inglis et al., 2003 [34]

When the mortality produced by the Nosema isolates was assessed in the laboratory using larvae that had stayed in environmental-like natural winter diapause conditions, variations in mortality between 0 and 55% were observed in larvae, and variations between 7 and 29% were observed in pupae; homogenization of the dead larvae revealed a large amount of Nosema spores. However, in the surviving adults, a large number of larvae were positive for the Nosema spores when they were analyzed under light microscopy using staining techniques and electron microscopy.

In these experiments, the larvae were reared in the winter diapause stage, and the temperature for Nosema was not optimal for the development of infection; in the field during the winter, the prediapause larvae migrate to the base of the corn stem just below the surface of the soil [5], and it is possible that Nosema can produce infection in natural conditions as the temperature inside the cane is higher than the external temperature. Phoofolo et al. [35] observed a similar behavior, and they reported that the O. pyrausta infection by Nosema is chronic. Although no immediate mortality is produced, the longevity and fecundity of the adults are reduced. Likewise, Phoofolo mentioned the possibility of a response to other factors of mortality that regulate the population dynamics of borers, such as low temperature, the host plant, and crowding, which have chronic effects on Nosema infection. Therefore, although the pest did not die directly because of microsporidia infection, the population size was eventually reduced.

This effect has been previously observed by Fuxa [32], when he evaluated the susceptibility of larvae in the first and third instars of six species of Lepidoptera, including D. saccharalis, to the microsporidium Vairimorpha necatrix. He described two routes of infection that resulted in mortality; one resulted from the chronic effects produced by the exposure of larvae to low doses of spores, which led to lethal septicemia (microsporidiosis) just before pupating, and the other route was due to the intake of a large number of spores, which apparently damaged the gut because of the introduction of a large number of spore polar filaments. In this study, Fuxa [32] concluded that this pathogen could be promising for the remaining five species, although not for D. saccharalis, as he had observed direct mortality by damage to the gut and indirect mortality by septicemia and because a high concentration of spores was necessary.

Inglis et al. [34] reported that one strain of Nosema (506), which had been previously isolated [5], can infect other species of insects in the Crambidae family, including O. nubilalis and D. crambidoides (Grote), but not other species of Noctuidae. However, they also observed that the infection can be transmitted transovarially, although at a very low frequency (a difference with other Nosema species that can be transmitted frequently in the vertical rout, such as the genus Ostrinia).

Solter et al. [36] researched the vertical and horizontal transmission of seven species of microsporidia, including two strains of Nosema sp. (isolated from their natural hosts in the field D. saecharalis and Eoreuma loftini); although five of these strains were transmitted at a low percentage horizontally and vertically, twoNosema strains did not behave similarly.

In assessing the infectivity percentage under laboratory conditions when applying high concentrations of spores, they observed very low mortality in the larvae of D. sacharalis and E. loftini (2 and 4%, resp.), although the same strains produced high mortality at low concentrations in the other species studied, including O. nubilalis. They explained the low mortality, even at high doses, as a function of low horizontal and vertical transmission because few infective or abnormal spores were produced.

In this case, it is important to clarify that they could obtain live infected larvae without mortality; however, they only measured mortality and did not measure other parameters, such as fecundity or larvae hatching in the next generation. As mentioned by Fuxa [32] and Inglis et al. [5], infection does not always lead to mortality, and occasionally the effects are observed long term and can be inferred from a reduction in fecundity and susceptibility to other stress situations.

Lastra and Gómez [16] implemented a system to produce natural enemies of D. saccharalis in CENICAÑA (Clombian Sugarcane Research Center, Colombia) and began their colony with larvae collected in the field. Their most significant result was the detection of one "protozoa," possibly Nosema, identified as the causal agent for the diminution in larvae production. They observed refractile spores in the fluids of the malpighi tubes or hemolymph using phase contrast microscopy in healthy adults. Macroscopically, the diseased larvae were dwarfed and white.

In addition, different pathogenic microorganisms have been isolated from Diatraea sp. in experimentally transmitted infections in the laboratory; however, there is little information about the natural presence of entomopathogens belonging to the microsporidia group in this species of borer. It is important to perform a systematic search for infected larvae, pupae, or adults of Diatraea, and studies need to include microscopy techniques to complement the mortality bioassays because infections could be asymptomatic. As many microsporidia do not produce insecticide activity quickly and because many species have complex life cycles that involve more than one host, very few attempts have been made to implement microsporidia and use them as agents of biological control.

NEMATODES

Pathogenic nematodes of the Heterorhabditidae and Steinernematidae families live in the soil where they are parasitic to certain soil-dwelling insects. The free-living third juvenile stage (infective juveniles, IJs) locates a suitable host and penetrates the host in different ways. Once inside the insect, the IJs initiate its development, the nutritional tract becomes functional, and the symbiotic bacterium Xenorhabdus or Photorhabdus is released through the anus and begins to multiply in the hemocoel, which kills the insect by septicemia and creates suitable conditions for the reproduction of the nematode. Nematodes feed on bacteria and the dead tissue of the host, and they pass through several generations until new IJs are produced, which emerge from the cadaver [37]. They have a ubiquitous distribution [37]; therefore, it is not unusual to find them in the soil of sugarcane plantations (Pizano et al., 1985 cited in Khan et al. [38, 39]).

However, there are few reports of nematodes infecting stem borers of the Diatraea genus in natural and experimental situations. In Costa Rica, several entomopathogenic species, including nematodes, have been isolated from D. tabernella [40]; however, the author does not mention the nematode species.

Khan et al. [38] reviewed the world bibliography on rice stem borers and found five reports of entomopathogenic nematodes associated with D. saccharalis, two reports of Steinernema (=Neoaplectana)glaseri in Brazil and three reports of Steinernema (=Neoaplectana) carpocapsae in the USA and Guadeloupe. However, in one study, they report that the aim of the research was the use of D. saccharalis to produce S. glaseri in controlled conditions because of its susceptibility to the nematode. Another study by Folegatti et al.[41] used entomopathogenic nematodes and D. saccharalis in laboratory conditions to produce S. carpocapsae in vivo, using larvae of the sugarcane borer as the host.

We believe that pathogenic nematodes have good potential for the biological control of Diatraea species because of their presence in sugarcane-producing regions (as mentioned above) and because the species S. feltiae, S. glaseri, S. rarum, Heterorhabditis heliothis, and H. bacteriophora have demonstrated high infectivity in D. saccharalis [42–45] (Table 4) as well as in E. loftini (some of them) [46, 47] in experimental studies. We consider that performing a systematic search for infected larvae or pupae of Diatraea could be helpful in finding new nematode strains and species adapted to local environmental conditions and pest species that could be used in the future.

Table 4: Entomopathogenic-nematode bioassays on Diatraea spp

Host specie	Pathogen	Bioassay	Result	Reference
	Steinernema feltiae [Neoaplectana carpocapsae]		100% mortality with 5000 nematodes/ larvae	
Diatraea saccharalis	Heterorhabditis heliothidis	Several concentrations for all species	100% mortality with 5000 nematodes/ larvae	Sosa et al., 1993 [42]
	S. glaseri		30% mortality with 5000 nematodes/ larvae	
Diatraea saccharalis	S. feltiae S. rarum Heterorhabditis bacteriophora	500 IJs/5 larvae in Petri dishes with two moistened filter papers disks	Adults and IJs produced. Larval mortality >90%	De Doucet et al., 1999 [43]
Diatraea saccharalis	H. bacteriophora JPM4 isolated from soil in Larvae MG, Brazil	Larvae were exposed to 25 5 IJs/100 L for 48 h in sand-filled Petri dishes	100% larval mortality within 5 days. LT50 2.1 d and LT95 4.4 d	Molina et al., 2007 [44]
Diatraea saccharalis	Heterorhabditis bacteriophora	500 mL/L applied to larvae	93% of larvae mortality	Aguila et al., 2008 [45]

Although we could not find any report on field trials, we are aware that there have been mass rearings of H. bacteriophora in Cuba since 1987 to control soil pests, such as D. saccharalis [45].

BACTERIA

Bacillus thuringiensis (Bt) is a Gram-positive bacterium that has been isolated from several sources, including soil, water, phylloplane, and insect cadavers. Bt produces various insecticidal crystal proteins during the onset of sporulation that are toxic to insects, acari, and nematodes [48]. The steps involved in the mode of action of the proteins after their ingestion are as follows: (a) solubilization of the crystals by the highly alkaline pH of the midgut, (b) activation of the proteins by proteases, (c) binding of the toxins to specific receptors located on the microvilli membrane of the midgut columnar epithelium cells, and (d) the insertion of the toxin into the membrane, which forms a pore and induces cell lysis. There are no reports from the American continent on the isolation of Bt strains from Diatraea spp. larvae cadavers; however, some Bt strains and pure proteins have been evaluated against these pests.

Bohorova et al. [49] evaluated Cry1Aa, Cry1Ab, Cry1Ac, Cry1B, Cry1C, Cry1D, Cry1E, and Cry1F Bt pure proteins against four species of Lepidoptera that are pests of maize, including D. saccharalis. The proteins were diluted in water and added to the diet at doses of 10 and 100 mg/mL of diet, and mortality was recorded after seven days. D. saccharalis was susceptible to Cry1B protein at 10 mg/g, and the LC_{50} was 113.6 mg/g of meridic diet (Table 5).

Table 5: Entomopathogenic-bacteria bioassays on Diatraea spp

Host species	Pathogen	Bioassay	Result	Reference
Diatraea saccharalis	Bacillus thuringiensis	Diet incorporated spores-crystals	LC50 113.6 mg/g diet	Bohorova et al., 1997 [49]
D.saccharalis	B. thuringiensis		LC50 33.21 g/mL	Rosas-Garcia et al., 2004 [50]
D.saccharalis	B. thuringiensis	Diet incorporated spores-crystals complex	LC50 13.06 g/L	Gitahy et al., 2007 [51]
D.saccharalis	Photorhabdustemperata	Cells injected	LD50 16.2 bacterial cells LT50 33.8 h	Carneiro et al., 2008 [52]
Diatraea grandiosella	Bacillus thuringiensis	Diet incorporated spores-crystals	LC505.2 mg/gdiet	Bohorovaetal.,1997[49]
Diatraeagrandiosella	Bacillusthuringiensis BiotrolBTB183-25NutriliteProductsIncorporated,KansasCity,Missouri,USA	Immersingstemseedlinginsporessuspensions:1.25×108,6.25×107,2.5×107and1.25×107	Meanmortalityon5-day-oldlarvae:32,31,25,15outof40 Controlmeanmortality0.01 Meanmortalityon10-day-oldlarvae:15,15,12and6outof20.	SikorowskiandDavis,1970[53]

Twelve Bt strains were evaluated by Rosas García et al. [50] at a dosages of 50 and 500 μg of total protein and spores per milliliter against 2-day-old D. saccharalis larvae. The strains used were HD1, HD2, HD9, HD29, HD37, HD59, HD133, HD137, and HD559, as well as the GM7, GM10, and GM34 native strains. The strains that killed more than 50% of larvae were selected to obtain the LC_{50}. Strains HD133, HD559, GM7, GM10, and GM34 were toxic; however, GM34 was the most toxic with an LC_{50} of 33.21 μg/mL (Table 5). PCR analysis was performed to determine the cry1 genes of the toxic strains: HD133 cryAa, cry1Ab, cry1C; HD559 and GM7cry1Aa, cry1Ab, and cry1B; GM10 cry1Aa, cry1Ab, cry1Ac, and cry1C; GM34 cry1Aa, cry1Ab, and cry1Ac.

Gitahy et al. [51] evaluated a spore-crystal complex in vitro from five native Bt strains (S48, S76, S90, S105, and S135) and HD1, which served as a positive control, on second instar larvae of D. saccharalis and mortality was recorded after 5 days. The strain S76 caused 100% mortality at 72 h, HD1 caused 69% mortality, and 3% of the other native strains caused mortality at 500 μg L^{-1} of the spore-crystal complex. The LC_{50} values of the S76 and HD1 strains were determined, and S76 was 11-fold more toxic (13.06 μg/L) than HD1 (143.88 μ/L) (Table5). The S76 strain carries cryAa, cry1Ab, cryAc, cry2Aa, and cry2Ab genes, which is similar to HD1.

There are other species of bacteria with potential insecticide activity; Carneiro et al. [52] evaluatedPhotorhabdus temperata, which is a bacterium associated with Heterorhabditis entomopathogenic nematodes. Cells were injected with a volume of 10 μL of phosphate-buffered saline directly into the hemocoel of fourth instar D. saccharalis larvae, and the LD_{50} was 16.2 bacterial cells with an LT_{50} of 33.8 h.

Sikorowski and Davis [53] determined the susceptibility of 5- and 10-day-old D. grandiosella larvae with a Bt commercial product (Biotrol BTB 183-25) using 1.25×10^8, 6.25×10^7, 2.5×10^7, and 1.25×10^7 spores per milliliter (Table 5) where two-inch stem seedlings were dipped for 30 min in spore suspensions. One 10-day-old or two 5-day-old larvae were allowed to feed on each stem. In addition, a known number of spores were placed or injected into 5-day-old stem seedlings at inch sections. Stems dipped in water were used as controls. Mortality was recorded after 48 h. With 10-day-old larvae, five replicates with 20 larvae for each treatment were used; for the

5-day-old larvae, two replicates with 40 larvae per treatment were used. With 5- and 10-day-old larvae at 1.25×10^8 spores per milliliter, a mean of 32 dead larvae out of 40 and 15 dead larvae out of 20 were recorded (Table 5), respectively. Thus, it was concluded that this species is highly susceptible toB. thuringiensis.

VIRUSES

Viruses that infect insects have received great attention as biological control agents because of their specificity on insect populations; they have little or no impact on the environment and are an ecological friendly alternative to chemical pesticides [58]. Entomopathogenic viruses are grouped in 33 genera within fifteen families [59]. However, only a few of these families, such as Baculoviridae, Poxviridae, and Reoviridae, have potential as biological control agents and have been successfully used in microbial control programs. A common characteristic among these entomopathogenic viruses is that the virions (infective unit) are occluded within a crystalline protein matrix to form an occlusion body (OB) [60], which is a unique characteristic of viruses that infect insects. In these entomopathogenic viruses, OBs have independently evolved as a protective mechanism to environmental factors, which gives the viruses a great advantage as biological control agents [61]. Baculoviruses (BVs) and entomopoxviruses (EPVs: subfamily Entomopoxvirinae) have a large double-stranded DNA genome, and cypoviruses (CPVs: family Reoviridae, genera Cypovirus) contain segmented double-stranded RNA viruses.

EPVs have been reported to infecting insects of a variety of orders, such as Coleoptera, Lepidoptera, Orthoptera, and Diptera. Some EPVs have two distinct OBs spheroids and spindles. The spheroids occlude virions, whereas the spindles do not [62]. The most abundant proteinaceous component of spheroids and spindles is proteins called spheroidin and fusolin, respectively [63–66]. EPV fusolin is an enhancing factor (EF) that increases BVs infection and has been characterized as a chitin-binding protein [67]. The fusolin mechanism of action is similar to that of Calcofluor, which facilitates BV infection by disrupting or preventing the formation of the peritrophic membrane [68]. EPVs are pathogenic but are scarcely virulent; infected larvae exhibit extreme longevity and take up to 70 days to die. However, EPVs have potential as biological control agents for pest insects where BVs have not been isolated [61]. Because of the activity of spindles, the EF of EPVs can be used as a synergistic agent to increase BV infectivity or generate genetically modified organisms, such as BVs and transgenic plants.

CPVs have been mainly isolated from Lepidoptera insects. OBs are dissolved in the midgut, and virions only infect epithelial cells; therefore, CPVs are very pathogenic but act slowly and frequently to produce chronic infections [69]. At this time, no commercial bioinsecticides based on CPVs have been developed. However, Caballero and Williams [61] suggest that their greatest potential as biological control agents is through inoculative or augmentative releases.

BVs predominantly infect insects within the Lepidoptera order, which includes important agricultural insect pests [70]. The Baculoviridae family includes two genera: nucleopolyhedrovirus (NPV), which forms large, polyhedric OBs where many enveloped virions are occluded [71], and Granulovirus (GV), which forms small, granular OBs that occlude only one enveloped virion each [72]. BVs are safe for humans and wildlife. Their specificity is usually very narrow and often is species specific. Because of their specificity and other characteristics, such as elevated virulence and pathogenicity, BVs are by far the most studied and extensively used as commercial biopesticides for the control of a variety of insect pests in many countries around the world [60, 61, 73].

Stem borers of the Lepidoptera order attack gramineous crops throughout the world [74, 75]. Diatraea stem borers (DSB) are widely distributed in the Americas and attack a wide variety of host plants, including maize and sugarcane [1]. Prediapause larvae of southwestern corn borers migrate to the base of the stalk of the corn plant below the soil surface to survive the winters [76]; cryptic habits make the chemical control of these insect pests difficult. Degaspari et al. [77] have argued that chemical control of D. saccharalis in Brazil is not economically feasible. To solve this problem, surveys to isolate endemic entomopathogens of stem borer populations in maize and sugarcane crops have been developed.

Inglis et al. [5] developed an exhaustive survey to isolate entomopathogens from the southern corn borer, D. grandiosella, and the southern corn stalk borer, D. crambidoides, and larvae in the diapause stage were collected from crops located in Mississippi and North Carolina. These authors did not observe OBs in any of the collected larvae and concluded that there are no naturally occurring viruses in these Diatraea populations [5]. According to Inglis et al. [5], Pavan and Ribeiro [55] mentioned that natural populations of the SCB in Brazil do not exhibit endemic viral pathogens.

Currently, there are only two records of endemic entomopathogenic viruses isolated from Diatraea spp. larvae. Pavan et al. [17] isolated a GV from the sugarcane borer (SCB), D. saccharalis, from sugarcane crops in the southern United States (Table 6). These authors developed bioassays with D. saccharalis third instar larvae (Table 6). External symptoms of GV-infected larvae were similar to those reported for other lepidopterous larvae, and the symptoms and ultrastructures were determined using electronic microscopy as well as replication of DsGV, which are typical of GVs. The LD_{50} for the third instar larvae was 42.3 OBs/larva (Table6) with 14.5 and 123.6 OBs/larva as the lower and upper limits at 95% probability, respectively [16]. These authors stated that this work was the first of a series of publications; however, there have been no additional studies produced to date, such as the molecular and biological characterization, of this GV, DsGV was introduced into Brazil [16], and Moscardi [78] mentioned that this virus is currently being applied at a small scale as an experimental product on sugarcane.

Table 6: Entomopathogenic-virus bioassays on Diatraea spp

Host species	Pathogen	Bioassay conditions	Result	Reference
D. grandiosella	Autographa californica MNPV (AcMNPV)	4 and 7 day-old larvae were fed strips of corn leaf soaked in different dilutions of virus.	DL50: 1.32 103 PIB*/ larvae (for 4-day-old larvae) DL50: 1.32 104 PIB/ larvae (for 7-day-old larvae).	Davis and Sikorowsy, 1978 [54]
D. saccharalis	Diatraea saccharalisGranulosis Virus (DsGV)	3rd instar larvae feeding individually on small artificial diet discs treated with 2.7 μL of virus dilutions.	LD50: 42 PIB/larva at 26°C LT50: From 29 to 63 for 107 and102 PIB/larvae, respectively.	Lastra and Gómez et al., 1983 [16]
D. saccharalis	Anticarsia gemmatalisMNPV(AgMNPV)	3rd instar larvae fed with artificial diet discs treated with 5 different doses of PIB through 20 serial passages.	LD50: From 7.9 105 to 5.3 102 for 1 and 20 serial passages, respectively.	Pavan and Ribeiro, 1989 [55]

D. saccharalis	AgMNPV wt AgMNPV-D10	3rd instar larvae individually fed small artificial diet discs containing 105 PIB of virus and kept individually at 10 different temperatures.	LT50: From 10 to 13 for 39 and 30°C, respectively From 9 to 29 for 39 and 22°C, respectively	Ribeiro and Pavan, 1994 [56]
D. saccharalis	Trichoplusia niMNPV(TnMNPV) wt TnMNPV-D11	3rd instar larvae fed individually with small artificial diet discs containing 105 PIB of virus and kept individually at 10 different temperatures.	LT50: From 9 to 16 for 39 and 24°C, respectively From 6 to 46 for 39 and 17°C, respectively	Ribeiro and Pavan, 1994 [56]
D. sacccharalis	AgMNPV	3rd instar larvae fed on formalin free diet disc (3 mm) inoculated with 2.7 μL of the viral solution from 10 clonal isolates of the AgMNPV-Ds20	LD50: From 5.3 × 103 PIB/larvae (7A genotypic variant) to 8 × 104 PIB/larvae (33B genotypic variant)	Ribeiro et al., 1997 [57]

*PIB: Polyhedral inclusion bodies.

Because of their high pathogenicity combined with a limited host range, Densoviruses (DNVs) have potential as effective insecticides [79]. Meynadier et al. [80] isolated a DNV of D. saccharalis (DsDNV) from the Guadeloupe sugarcane borer (Table 6). Kouassi et al. [81] tested the pathogenicity of DsDNV on its host. These authors observed that the infected larvae exhibited infection symptoms from the fourth day postinfection, such as anorexia and lethargy followed by flaccidity and inhibition of molting and metamorphosis. Larvae became paralyzed and stopped feeding after 7 days. The cumulative mortality of infected larvae increased significantly and reached 60% after 12 days and 100% at 21 days postinfection [81]. Although DNVs have no potential for large scale use as a biological control agent because they have no OBs and are related to vertebrate pathogenic viruses, the genes involved with anorexia and paralysis could be used to produce transgenic BVs or plants.

Most BVs have a limited host range and may be species specific in some cases, although there are several NPVs, such as Autographa californica MNPV, Anagrapha falcifera MNPV, and Anticarsia gemmatalis MNPV, which have broader host ranges within the Lepidoptera order [60]. Because of the scarcity of reports on natural isolates of entomopathogenic viruses from Diatraea spp. populations, cross-infectivity of AgMNPV, TnMNPV, and AgMNPV has been evaluated in D. saccharalis and D. grandiosella larvae (Table 6).

Although only two reports of entomopathogenic viruses isolated from Diatraea stem borers exist in the Americas, entomopathogenic viruses, such as NPVs and GVs, have been reported to occur in Africa and Asia in cereal stem borers within the Crambidae family, including maize and sugarcane borers such as Chilo sp. [82], Chilo sacchariphagus [83, 84], Ch. infuscatellus [84, 85], Ch. partellus [86], and Eldana saccharina [87]. As the Mexican territory is the origin of some Diatraea stem borer species, we hypothesize that a great diversity of pathogenic viruses in Diatraea populations that attack sugarcane crops exists in Mexico.

ACKNOWLEDGMENTS

This study was financed in part by the project "Selección de enemigos naturales de barrenadores de la caña de azúcar de Colima y Morelos con potencial como agentes de control biológico (SEP-PROMEP)," of Mexico. The authors also thank Mrs. Ingrid Masher for reviewing the paper and for editorial assistance.

REFERENCES

1. L. A. Rodriguez-Del-Bosque and J. W. Smith Jr., "Biological control of maize and sugarcane stemborers in Mexico: a review," Insect Science and Its Application, vol. 17, no. 3-4, pp. 305–314, 1997.
2. G. Serra and E. Trumper, "Estimación de incidencia de daños provocados por larvas de Diatraea saccharalis (Lepidoptera: Crambidae) en tallos de maíz mediante evaluación de signos externos de infestación," Agrosciencia, vol. 23, no. 1, pp. 1–7, 2006.
3. M. A. P. De Oliveira, E. J. Marques, V. Wanderley-Teixeira, and R. Barros, "Effect of Beauveria bassiana(Bals.)Vuill. and Metarhizium anisopliae (Metsch.) Sorok. on biological characteristics of Diatraea saccharalis F. (Lepidoptera: Crambidae)," Acta Scientiarum, vol. 30, no. 2, pp. 220–224, 2008.
4. R. Reyes, "Sorghum stem borers in Central and South America," in Proceedings of the International Workshop on Sorghum Stem Borers, pp. 49–60, ICRISAT Center, Hyderabad, India, 1987.
5. G. D. Inglis, A. M. Lawrence, and F. M. Davis, "Pathogens associated with southwestern corn borers and southern corn stalk borers (Lepidoptera: Crambidae)," Journal of Economic Entomology, vol. 93, no. 6, pp. 1619–1626, 2000.
6. M. D. R. T. De Freitas, E. L. Da Silva, A. D. L. Mendonça et al., "The biology of Diatraea flavipennella(Lepidoptera: Crambidae) reared under laboratory conditions," Florida Entomologist, vol. 90, no. 2, pp. 309–313, 2007. View at Publisher · View at Google Scholar ·
7. R. H. R. Destéfano, S. A. L. Destéfano, and C. L. Messias, "Detection of Metarhizium anisopliae var.anisopliae within infected sugarcane borer Diatraea saccharalis (Lepidoptera, Pyralidae) using specific primers," Genetics and Molecular Biology, vol. 27, no. 2, pp. 245–252, 2004.

8. A. E. Knutson and F. E. Gilstrap, "Seasonal occurrence of Beauveria bassiana in the southwestern corn borer (Lepidoptera: Pyralidae) in the Texas High Plains," Journal of the Kansas Entomological Society, vol. 63, no. 2, pp. 243–225, 1990.

9. K. Zambrano, M. Dávila, and M. A. Castillo, "Detección de fragmentos de AND de hongos y su posible relación con la síntesis de proteínas de actividad entomopatógena," Revista de la Facultad De Agronomía (LUZ), vol. 19, pp. 185–193, 2002.

10. S. B. Alves, L. S. Rossi, R. B. Lopes, M. A. Tamai, and R. M. Pereira, "Beauveria bassiana yeast phase on agar medium and its pathogenicity against Diatraea saccharalis (Lepidoptera: Crambidae) andTetranychus urticae (Acari: Tetranychidae)," Journal of Invertebrate Pathology, vol. 81, no. 2, pp. 70–77, 2002.

11. M. E. Estrada, M. Romero, M. J. Rivero, and F. Barroso, "Natural presence of Beauveria bassiana(Balsamo) Vuillemin in the sugar cane (Saccharum sp. hybrid) in Cuba," Revista Iberoamericana de Micologia, vol. 21, no. 1, pp. 42–43, 2004.

12. M. G. Yasen de Romero, A. R. Salvatore, G. López, and E. Willink, "Presencia natural de hongos hyphomycetes en larvas invernantes de Diatraea ssaccharalis F. en caña de azúcar en Tucumán, Argentina," Revista Industrial y Agrícola de Tucumán, vol. 85, no. 2, pp. 39–42, 2008.

13. C. A. Angel-Sahagún, R. Lezama-Gutiérrez, J. Molina-Ochoa et al., "Susceptibility of biological stages of the horn fly, Haematobia irritans, to entomopathogenic fungi (Hyphomycetes)," Journal of Insect Science (Online), vol. 5, no. 50, pp. 1–8, 2005.

14. R. A. Humber, K. S. Hansen, and M. M. Wheeler, Catalog of Species. ARS Collection of Entomopathogenic Fungal Cultures, USDA-ARS Biological Integrated Pest Management Research, Ithaca, NY, USA, 2009.

15. V. M. Hernández Velázquez and R. Lezama Gutiérrez, "Uso de entomopatógenos para el control biológico de barrenadores del tallo," in Taller Internacional Sobre Barrenadores Del Tallo De Caña De Azúcar, L. A. Rodríguez del Bosque, G. Vejar, and E. Cortéz, Eds., pp. 37–45, Sociedad Mexicana de Control Biológico, Los Mochis, Sinaloa, Mexico, Noviembre de 2004.

16. B. Lastra and L. A. Gómez, "Cría y producción masiva de insectos en un programa de control biológico en caña de azúcar," in 1er Curso Taller Internacional de Control Biológico, pp. 335–340, Memorias, Bogotá, Colombia.

17. O. H. O. Pavan, D. G. Boucias, L. C. Almeida, J. O. Gaspar, P. S. M. Botelho, and N. Degaspari, "Granulosis Virus of Diatraea saccharalis (DsGV): pathogenicity, replication and ultrastructure," inProceedings of the International Congress of the International Society of Sugar Cane Technologists (ISSCT ‹83), vol. 2, pp. 644–659, La Havana, Cuba, 1983.

18. G. Meynadier, P. F. Galichet, J. C. Veyrunes, and A. Amargier, "Mise en évidence d›une densonucléose chez Diatraea saccharalis [Lep.: Pyralidae]," Entomophaga, vol. 22, no. 1, pp. 115–120, 1977. · ·

19. A. Fonseca-González, G. Peña-Chora, A. Trejo-Loyo, L. Lina-García, L. A. Rodríguez-Del Bosque, and V. M. Hernández-Velázquez, "Virulencia de seis aislados de Bacillus thuringiensis nativos delestado de Morelos y evaluados sobre Diatraea magnifactella," in Memoria IV Congreso Nacional de Control Biológico, pp. 31–34, Sociedad Mexicana de Control Biológico, Monterrey, Nuevo León, México, Noviembre 2011.

20. S. C. M. Leal, D. J. Bertioli, T. M. Butt, and J. F. Peberdy, "Characterization of isolates of the entomopathogenic fungus Metarhizium anisopliae by RAPD-PCR," Mycological Research, vol. 98, no. 9, pp. 1077–1081, 1994.

21. S. Ali, Z. Huang, and S. Ren, "Media composition influences on growth, enzyme activity, and virulence of the entomopathogen hyphomycete Isaria fumosoroseus," Entomologia Experimentalis et Applicata, vol. 131, no. 1, pp. 30–38, 2009. · ·

22. J. C. Legaspi, T. J. Poprawski, and B. C. Legaspi Jr., "Laboratory and field evaluation of Beauveria bassiana against sugarcane stalkborers (lepidoptera: pyralidae) in the Lower Rio Grande Valley of Texas," Journal of Economic Entomology, vol. 93, no. 1, pp. 54–59, 2000.

23. L. F. Solter and J. V. Maddox, "Microsporidia as classical biological control agents: research and regulatory issues," Phytoprotection, vol. 79, no. 4, pp. 75–80, 1998.

24. N. Corradi and P. J. Keeling, "Microsporidia: a journey through radical taxonomical revisions," Fungal Biology Reviews, vol. 23, no. 1-2, pp. 1–8, 2009.

25. J. Weiser, "Microsporidia and the society for invertebrate pathology: a personal point of view," Journal of Invertebrate Pathology, vol. 89, no. 1, pp. 12–18, 2005.

26. J. A. Arcas, B. M. Díaz, and R. E. Lecuona, "Bioinsecticidal activity of conidia and dry mycelium preparations of two isolates of Beauveria bassiana against the sugarcane borer Diatraea saccharalis,"Journal of Biotechnology, vol. 67, no. 2-3, pp. 151–158, 1999.

27. J. C. Legaspi Jr., B. C. Legaspi, and R. R. Saldaña, "Evaluation of Steinernema riobravis (Nematoda: Steinernematidae) against the Mexican rice borer (Lepidoptera: Pyralidae)," Journal of Entomological Science, vol. 35, no. 2, pp. 141–149, 2000.

28. E. J. Marques, S. B. Alves, and I. M. R. Marques, "Virulencia de Beauveria bassiana (Bals.) Vuill. aDiatraea saccharalis (F.) (Lepidoptera: Crambidae) após armazenamiento de conídios em baixa temperatura," Anais da Sociedade Entomológica do Brasil, vol. 29, no. 2, pp. 303–307, 2000.

29. S. B. Alves, L. S. Rossi, R. B. Lopes, M. A. Tamai, and R. M. Pereira, "Beauveria bassiana yeast phase on agar medium and its pathogenicity against Diatraea saccharalis (Lepidoptera: Crambidae) andTetranychus urticae (Acari: Tetranychidae)," Journal of Invertebrate Pathology, vol. 81, no. 2, pp. 70–77, 2002.

30. I. M. Wenzel, F. H. C. Gionetti, and J. E. M. Almeida, "Patogenicidade do isolade IBCB66 de Beauveria bassiana s broca da caña-de-acúcar Diatraea saccharalis em condicoes de laboratório," Arquivos Do Instituto Biologico, vol. 72, no. 2, pp. 259–261, 2006.

31. C. E. Lange, "Niveles de esporulación experimentales y naturales de Nosema locustae (Microsporidia) en especies de tucuras y langostas (Orthoptera: Acridoidea) de la Argentina," Revista de la Sociedad Entomológica Argentina, vol. 62, no. 1-2, pp. 15–22, 2003.

32. J. R. Fuxa, "Susceptibility of Lepidopterous pests to two types of mortality caused by the microsporidiumVairimorpha necatrix," Journal of Economic Entomology, vol. 74, no. 1, pp. 99–102, 1981.

33. L. F. Solter and J. V. Maddox, "Microsporidia as classical biological control agents: research and regulatory issues," Phytoprotection, vol. 79, no. 4, pp. 75–80, 1998.
34. G. D. Inglis, A. M. Lawrence, and F. M. Davis, "Impact of a novel species of Nosema on the Southwestern corn borers (Lepidoptera: Cambridae)," Journal of Economic Entomology, vol. 96, no. 1, pp. 12–20, 2003.
35. M. W. Phoofolo, J. J. Obrycki, and L. C. Lewis, "Quantitative assessment of biotic mortality factors of the european corn borer (lepidoptera: crambidae) in field corn," Journal of Economic Entomology, vol. 94, no. 3, pp. 617–622, 2001.
36. L. F. Solter, J. V. Maddox, and C. R. Vossbrinck, "Physiological host specificity: a model using the European corn borer, Ostrinia nubilalis (Hübner) (Lepidoptera: Crambidae) and microsporidia of row crop and other stalk-boring hosts," Journal of Invertebrate Pathology, vol. 90, no. 2, pp. 127–130, 2005.
37. G. O. Poinar Jr., "Taxonomy and biology of Steinernematidae and Heterorhabditidae," inEntomopatogenic Nematodes in Biological Control, R. Gaugler and H. K. Kaya, Eds., pp. 23–61, CRC Press, 1990.
38. Z. R. Khan, J. A. Litsinger, A. T. Barrion, F. F. Villanueva, N. J. Fernandez, and L. D. Taylo, World Bibliography of Rice Stem Borers 1794–1990, International Rice Research Institute, 1991.
39. L. Dasrat, "Discovery of an indigenous entomopathogenic nematode and its pathogenicity to two sugar cane stem borers in Guyan,"
40. F. Badilla, "Un programa exitoso de control biológico de insectos plaga de la caña de azúcar en Costa Rica," Manejo Integrado de Plagas y Agroecología (Costa Rica), vol. 64, pp. 77–87, 2002.
41. M. E. Folegatti, S. Batista, P. R. Kawai, and P. S. Botelho, "Nova metodologia para produção in vivo deNeoaplectana carpocapsae Weiser," Nematologia Brasileira, vol. 12, pp. 76–83, 1988.
42. O. Sosa Jr., D. G. Hall, and W. J. Schroeder, "Mortality of sugarcane borer (Lepidoptera: Pyralidae) treated with entomopathogenic nematodes in field and laboratory trials," Journal of the American Society of Sugar Cane Technology, vol. 13, pp. 18–21, 1993.

43. M. M. A. De Doucet, M. A. Bertolotti, A. L. Giayetto, and M. B. Miranda, "Host Range, Specificity, and Virulence of Steinernema feltiae, Steinernema rarum, and Heterorhabditis bacteriophora(Steinernematidae and Heterorhabditidae) from Argentina," Journal of Invertebrate Pathology, vol. 73, no. 3, pp. 237–242, 1999.

44. J. P. M. Acevedo, R. I. Samuels, I. R. Machado, and C. Dolinski, "Interactions between isolates of the entomopathogenic fungus Metarhizium anisopliae and the entomopathogenic nematodeHeterorhabditis bacteriophora JPM4 during infection of the sugar cane borer Diatraea saccharalis(Lepidoptera: Pyralidae)," Journal of Invertebrate Pathology, vol. 96, no. 2, pp. 187–192, 2007.

45. P. J. Aguila, M. Vidal, M. González, L. A. Rodríguez, and E. Mesa, "Criterios ecológicos en el manejo de plagas de la caña de azúcar y cultivos varios una opción para lograr alimentos sanos," Agroecología, vol. 3, pp. 51–53, 2008.

46. D. R. Ring and H. W. Browning, "Evaluation of entomopathogenic nematodes against the Mexican rice borer (Lepidoptera: Pyralidae)," Journal of Nematology, vol. 22, no. 3, pp. 420–422, 1990.

47. J. C. Legaspi Jr., B. C. Legaspi, and R. R. Saldaña, "Evaluation of Steinernema riobravis (Nematoda: Steinernematidae) against the Mexican rice borer (Lepidoptera: Pyralidae)," Journal of Entomological Science, vol. 35, no. 2, pp. 141–149, 2000.

48. E. Schnepf, N. Crickmore, J. Van Rie et al., "Bacillus thuringiensis and its pesticidal crystal proteins,"Microbiology and Molecular Biology Reviews, vol. 62, no. 3, pp. 775–806, 1998.

49. N. Bohorova, M. Cabrera, C. Abarca et al., "Susceptibility of four tropical lepidopteran Maize pests toBacillus thuringiensis CryI-type insecticidal toxins," Journal of Economic Entomology, vol. 90, no. 2, pp. 412–415, 1997.

50. N. M. Rosas-García, B. Pereyra-Alférez, K. A. Niño, L. J. Galn-Wong, and L. H. Morales-Ramos, "Novel toxicity of native and HD Bacillus thuringiensis strains against to the sugarcane borer Diatraea saccharalis," BioControl, vol. 49, no. 4, pp. 455–465, 2004.

51. P. D. M. Gitahy, M. T. de Souza, R. G. Monnerat, E. D. B. Arrigoni, and J. I. Baldani, "A BrazilianBacillus thuringiensis strain highly active to sugarcane borer Diatraea saccharalis (Lepidoptera: Crambidae)," Brazilian Journal of Microbiology, vol. 38, no. 3, pp. 531–537, 2007.
52. C. N. B. Carneiro, R. A. DaMatta, R. I. Samuels, and C. P. Silva, "Effects of entomopathogenic bacteriumPhotorhabdus temperata infection on the intestinal microbiota of the sugarcane stalk borer Diatraea saccharalis (Lepidoptera: Crambidae)," Journal of Invertebrate Pathology, vol. 99, no. 1, pp. 87–91, 2008.
53. P. Sikorowski and F. M. Davis, "Susceptibility of larvae of the southwestern corn borer, Diatraea grandiosella, to Bacillus thuringiensis," Journal of Invertebrate Pathology, vol. 15, no. 1, pp. 131–132, 1970.
54. F. M. Davis and P. P. Sikorowsy, "Susceptibility of the Southwestern corn borer, Diatraea grandiosellaDyar (Lepidoptera: Pyralidae) to the baculovirus of Autographa californica," Journal of the Kansas Entomological Society, vol. 51, no. 1, pp. 11–13, 1978.
55. O. H. O. Pavan and H. C. T. Ribeiro, "Selection of a baculovirus strain with a bivalent insecticidal activity," Memórias do Insttituto Oswaldo Cruz, vol. 84, supplement 3, pp. 63–65, 1989.
56. H. C. T. Ribeiro and O. H. O. Pavan, "Effect of temperature on the development of baculoviruses,"Journal of Applied Entomology, vol. 118, pp. 316–320, 1994.
57. H. C. T. Ribeiro, O. H. O. Pavan, and A. R. Muotri, "Comparative susceptibility of two different hosts to genotypic variants of the Anticarsia gemmatalis nuclear polyhedrosis virus," Entomologia Experimentalis et Applicata, vol. 83, no. 2, pp. 233–237, 1997.
58. S. R. Palli, T. R. Ladd, W. L. Tomkins et al., "Choristoneura fumiferana entomopoxvirus prevents metamorphosis and modulates juvenile hormone and ecdysteroid titers," Insect Biochemistry and Molecular Biology, vol. 30, no. 8-9, pp. 869–876, 2000.
59. C. M. Fauquet, M. A. Mayo, J. Maniloff, U. Desselberger, and L. A. Ball, Eds., Virus Taxonomy, Eighth Report of the International Committee on Taxonomy of Viruses, Elsevier, San Diego, Calif, USA, 2005.

60. J. S. Cory and H. F. Evans, "Viruses," in Field Manual ofTechniques in Invertebrate Pathology, L. A. Lacey and H. K. Kaya, Eds., pp. 149–174, Springer, Dordrecht, The Netherlands, 2nd edition, 2007.

61. P. Caballero andT. Williams, "Virus Entomopatógenos," in Control Biológico de Plagas Agrícolas, J. A. Jacas and A. Urbaneja, Eds., pp. 121–135, PHYTOMA-España, Navarra, España, 2008.

62. W. Mitsuhashi, H. Kawakita, R. Murakami et al., "Spindles of an entomopoxvirus facilitate its infection of the host insect by disrupting the peritrophic membrane," Journal of Virology, vol. 81, no. 8, pp. 4235–4243, 2007.

63. M. Bergoin, J. C. Veyrunes, and R. Scalla, "Isolation and amino acid composition of the inclusions ofMelolontha melolontha poxvirus," Virology, vol. 40, no. 3, pp. 760–763, 1970.

64. R. L. Hall and R. W. Moyer, "Identification, cloning, and sequencing of a fragment of Amsacta mooreientomopoxvirus DNA containing the spheroidin gene and three vaccinia virus-related open reading frames," Journal of Virology, vol. 65, no. 12, pp. 6516–6527, 1991.

65. D. Dall, A. Sriskantha, A. Vera, J. Lai-Fook, and T. Symonds, "A gene encoding a highly expressed spindle body protein of Heliothis armigera entomopoxvirus," Journal of General Virology, vol. 74, no. 9, pp. 1811–1818, 1993.

66. T. Hayakawa, J. Xu, and T. Hukuhara, "Cloning and sequencing of the gene for an enhancing factor from Pseudaletia separata entomopoxvirus," Gene, vol. 177, no. 1-2, pp. 269–270, 1996.

67. Z. Li, C. Li, K. Yang et al., "Characterization of a chitin-binding protein GP37 of Spodoptera lituramulticapsid nucleopolyhedrovirus," Virus Research, vol. 96, no. 1-2, pp. 113–122, 2003.

68. P. Wang and R. R. Granados, "Calcofluor disrupts the midgut defense system in insects," Insect Biochemistry and Molecular Biology, vol. 30, no. 2, pp. 135–143, 2000.

69. T. Hukuhara and J. R. Bonami, "Reoviridae," in Atlas of Invertebrate Viruses, J. R. Adams and J. R. Bonami, Eds., pp. 393–434, CRC Press, Boca Raton, Fla, USA, 1991.

70. G. W. Blissard, B. Black, N. Crook, B. A. Keddie, R. Possee, et al., "Family Baculoviridae," in Virus Taxonomy: Seventh Report of the International Committee on Taxonomy of Virus, M. H. V. van Regenmoertel, C. M. Fauquet, D. H. L. Bishop et al., et al., Eds., pp. 195–202, Academic Press, San Diego, Calif, USA, 2000.

71. G. F. Rhormann, "Nuclear polyhedrosis viruses," in Encyclopedia of Virology, R. G. Webster and A. Granoff, Eds., pp. 146–152, Academic Press, London, UK, 2nd edition, 1999.

72. D. Winstanley and D. O. ‹Reilly, "Granuloviruses," in Encyclopedia of Virology, R. G. Webster and A. Granoff, Eds., pp. 140–146, Academic Press, London, UK, 2nd edition, 1999.

73. B. Szewczyk, L. Hoyos-Carvajal, M. Paluszek, I. Skrzecz, and M. Lobo De Souza, "Baculoviruses—re-emerging biopesticides," Biotechnology Advances, vol. 24, no. 2, pp. 143–160, 2006.

74. W. F. Jepson, A Critical Review of the World Literature of the Lepidopterous Stalk Borers of Graminaceous Crops, Commonwealth Institute of Entomology, London, UK, 1954.

75. J. W. Smith Jr., R. N. Wiedenmann, and W. A. Overholt, Parasites of Lepidopteran Stemborers of Tropical Gramineous Plants, ICIPE Science Press, Nairobi, Kenya, 1993.

76. G. M. Chippendale, "The southwestern corn borer, Diatraea grandiosella, case history of an invading insect," Research Bulletin 1031, Missouri Agricultural Experiment Station, Columbia, Mo, USA, 1979.

77. N. Degaspari, P. S. M. Botelho, and N. Macedo, "Controle quimico da Diatraea saccharalis em cana-de-acucar na região Centro Sul do Brasil," Boletim Técnico. IAA/PLANALSUCAR, vol. 3, pp. 1–16, 1981.

78. F. Moscardi, "Use of viruses for pest control in Brazil: the case of the nuclear polyhedrosis virus of the soybean caterpillar, Anticarsia gemmatalis," Memórias do Insttituto Oswaldo Cruz, vol. 84, pp. 51–56, 1989.

79. G. Fédière, "Epidemiology and pathology of Densovirinae," in Parvovir Uses. From Molecular Biology to Pathology and Therapeutic Uses, S. Faisst and J. Rommelaere, Eds., pp. 1–11, Karger, Basel, Switzerland, 2000.

80. G. Meynadier, P. F. Galichet, J. C. Veyrunes, and A. Amargier, "Mise en évidence d›une densonucléose chez Diatraea saccharalis [Lep.: Pyralidae]," Entomophaga, vol. 22, no. 1, pp. 115–120, 1977.

81. N. Kouassi, J. X. Peng, Y. Li, C. Cavallaro, J. C. Veyrunes, and M. Bergoin, "Pathogenicity of Diatraea saccharalis densovirus to host insets and characterization of its viral genome," Virologica Sinica, vol. 22, no. 1, pp. 53–60, 2007.

82. E. P. Steinhaus and G. A. Marsh, "Report of diagnosis of disease insects, 1951–1961," Hilgardia, vol. 33, pp. 349–490, 1962.

83. U. K. Metha and H. David, "A granulosis virus disease of sugarcane internode borer," Madras Agricultural Journal, vol. 67, pp. 616–619, 1970.

84. S. Easwaramoorthy and S. Jayaraj, "Survey of granulosis virus infection in sugarcane borers, Chilo infuscatellus Snellen and C. sacchariphagus indicus (Kapur) in India," Tropical Pest Management, vol. 3, pp. 200–201, 1987.

85. S. Easwaramoorthy and H. David, "A granulosis virus of sugarcane shoot borer, Chilo infuscatellus Snell. (Lepidoptera: Crambidae)," Current Science, vol. 48, pp. 685–686, 1979.

86. S. Sethuraman and K. Narayanan, "Biological activity of Nucleopolyhedrovirus isolated from Chilo partellus (Swinhoe) (Lepidoptera: Pyralidae) in India," Asian Journal of Experimental Biological Sciences, vol. 1, pp. 325–330, 2010.

87. A. J. Cherry, C. J. Lomer, D. Djegui, and F. Schuethess, "Pathogen incidence and their potential as microbial control agents in IPM of maize stem borers in West Africa," BioControl, vol. 44, no. 3, pp. 301–327, 1999.

Chapter 8

Isolation of Entomopathogenic Nematodes and Control of Phyllophaga Vetula Horn in Oaxaca, Mexico

Girón-Pablo S., Ruiz-Vega J., Pérez-Pacheco R., |Sánchez-García J. A, and Aquino-Bolaños T

Center for Interdisciplinary Research and Integral Regional Development (CIIDIR), National Polytechnical Institute (IPN), Oaxaca Unit, Mexico

ABSTRACT

This study aimed to isolate native entomopathogenic nematodes (EPN) in the Central Valleys of Oaxaca and to determine their potential for control of white grub (Phyllophaga vetula Horn). Fifty-five (55) soil samples were collected in 13 communities in the period August to October 2008 and 29.1% of these were found positive for EPN. Five isolates were selected for their apparent pathogenicity to third instar Galleria mellonella; two of the genus Steinernema and three

Heterorhabditis. After evaluating these five isolates for control of white grubs, the Heterorhabditidae EPN presented the lowest values for both lethal dosages and lethal times. Using polymerase chain reaction (PCR) procedures, the species Heterorhabditis mexicana, Steinernema carpocapsae and Steinernema feltiae were identified. We concluded that H. mexicana was the most effective isolate for control of P. vetula larvae.

INTRODUCTION

In the Central Valleys of Oaxaca Region, some 110,000 ha are planted with maize and Coleoptera larvae (white grubs) can cause severe plant damage. There are no available data to evaluate the extent of the problem, but severely infested plots decrease crop yields more than 50% (400 to 700 kg ha-1), mostly because of reduction in plant stand (Ruiz et al., 2003). Bravo (2003) found grain yield increases of 300 to 375 kg ha-1 when an insecticide was applied to plots infested with white grubs. Villalobos (1992) estimated maize yield reductions in the order of 400 to 1300 kg ha-1 in Central Mexico. In Oaxaca, most of the grain produced is for self-consumption and since it is a low-input agriculture, the farmers require a low-cost, yet effective method to control white grubs.

Entomo pathogenic nematodes (EPN) are obligate parasites in nature, which gives them the possibility of being used as bio-control agents and therefore represent a good alternative to chemical insecticides (Kaya and Gaugler, 1993); the third stage (infective juvenile) emerges from the host's body and can remain for months in the soil until it finds a new host (Kaya and Koppenhofer, 1996). Their hosts are larvae known as "white grubs" (Coleoptera: Scarabeidae), which are important pests of various crops, including maize and root crops (Koppenhofer, 2007).

It is recommended that native species are isolated, thereby increasing the chances of successful control; however, their effectiveness also depends on environmental factors such as soil moisture (Koppenhofer et al., 1995), temperature (Grewal et al., 1994), soil texture, and bulk density (Gruner et al., 2007).

The use of biological control agents cannot be generalized, but experimentation is required to determine which kind of NE is most suitable for controlling a certain species of white grubs (Koppenhofer, 2007). Therefore, when a new NE is isolated from field samples, laboratory assessments of ineffectiveness and control efficacy have to be done before applying it in the field (Wang and Grewal, 2002).

Also, it is important to know the host range and lethal dosages of the EPN, as well as their interactions with pest populations. This knowledge is necessary to use them effectively as an alternative to synthetic pesticides (Jackson, 2003).

The EPN of the families Heterorhabditidae and Steinernematidae have great potential as biological control agents of many insect pests (Griffin et al., 2005). These nematodes possess many desirable attributes, including wide host range, safety for non-target organisms and the environment, exemption from registration in many countries, easy mass production and application, ability to find the pest, and potential to persist in the environment, among others (Kaya and Gaugler, 1993).

The present study was aimed to isolate EPN from soil samples collected in 13 locations of the Central Valleys of Oaxaca Region, Mexico. The isolates were evaluated for their potential to control white grubs (*Phyllophaga vetula* Horn), and Probit Analysis (SAS, 2002) was used to determine the lethal dosages and times to control this soil dwelling pest.

MATERIALS AND METHODS

Soil Sampling

To detect and isolate local EPN species, soil sampling was carried out in the Central Valleys of Oaxaca Region, which has a mean annual temperature of 22.0°C and 650 mm rainfall, and is located at altitudes between 1500 to 1600 m. The sampling period was during the months of August to October, when the maize crop was in the reproductive stages and was actively growing because of adequate soil moisture, which increased the chances of finding EPN.

The presence of white grubs in the previous growing seasons was one of the criteria for choosing the sampling sites, as well as visible crop symptoms indicating the presence of the pest; theoretically, where there is a specific pest, there is also a natural enemy, which makes it an intentional sampling process (Azoríny and Sánchez, 1986).

Using a straight shovel to reach a depth between 10 and 30 cm, to discard the drier top layer, one sample (~1 kg of soil) per site was taken at about 10 cm from the base of the corn stalk; after homogenization in the shade to avoid temperature increases and solar radiation, which could affect the nematodes present in the sample, they were placed in a plastic bag to prevent moisture loss, and stored in a cool chest. After taking each sample, the blade was washed with water to prevent possible mixing of nematodes between sampling sites. A total of 55 sites were sampled.

Once in the laboratory, each 1 kg soil sample was placed in a plastic tray (20 × 15 × 7 cm) and ten late instar *Galleria mellonella* larvae were applied to each one, afterwards the tray was covered with aluminum foil to prevent moisture losses, but small holes were made to allow for gas exchange.

Every two days, over a period of two weeks, the trays were inspected to determine whether there were dead larvae; these were placed in White Traps (White, 1927) and when there was emergence of infective juveniles (IJ's), they were harvested and stored in properly labeled culture flasks at an average temperature of 10°C.

Experiment 1: Verification of Entomopatogenicity

An experiment to verify the entomopathogenicity of the isolates was carried out. Petri dishes were lined with filter paper and the 27 isolates were separately applied in an aqueous suspension at 300 IJ's per dish. Afterwards, 10 *G. mellonella* larvae were placed in each dish, which were placed inside plastic bags to prevent desiccation of the nematodes. After five days, the percentage of larval mortality was determined and each larva was placed in a White trap to verify their ability to produce IJ's in adequate numbers. This experiment was laid out in a completely randomized design and replicated four times.

Experiment 2: Lethal Dosages and Times for Selected Isolates

Using 1/3 of the most effective isolates detected in the bioassay with *G. mellonella*, five dosages: 50, 100, 200, 500 and 1000 IJ's per larva were evaluated. Also, a blank treatment without application of IJ's was included. There were 10 cups per treatment and the experimental design was a randomized one with four replicates. The plastic cups (100 ml) were previously filled with dry, sterilized soil to eliminate the influence of other pathogens. The soil used was loamy sand and was maintained at 80 ± 5% field capacity (FC). The water content at FC of this soil was 17.3%. Thus, to maintain the soil at 80% FC, 13.8 ml of distilled water was added initially. Every three days the cups were weighed and more water was added as needed to reach 80% FC again.

Each day after water was added one *P. vetula* larva was placed in the soil surface of each cup; after a waiting period of 30 min, non-active larvae were discarded, and the five different dosages of nematodes were applied in an aqueous suspension (200 μL). In order to avoid a rapid loss of moisture, the cups were covered with aluminum foil and a small hole was made with a toothpick.

Every other day, for 16 days, the cups were opened to determine the number of dead larvae. To determine whether the larva had actually been killed by a nematode infection, they were placed in White traps. Using the data from the tenth day after the start of the experiment, the lethal doses were determined, whereas all data collected during the duration of the experiment were used to determine the lethal times; both indicators were established for the control of 50 and 95% of the grub population.

The experiment was established under a completely randomized design and four replicates. Before data analysis, the mortality percentages were transformed with the arcsine square root procedure. The analysis of variance and Tukey›s test ($\alpha = 0.05$) for the separation of means was carried out with the Statistical Analysis System package (SAS, 2002).

Determination of the Species of Nematodes

The following procedure, the selected nematodes were identified by polymerase chain reaction (PCR) techniques (Ellis et al., 1986; Zhou et al., 1998) in the Nematology Laboratory of The University of California, Davis.

DNA Extractions

Individual adult nematodes were dissected from infected *G. mellonella* cadavers, cut in half and placed in a mix of 20 µl 'Chelex 100' resin (5%, Bio-Rad, Hercules, CA), and 1 µl proteinase K solution (20 mg/ml; Sigma, St. Louis, Missouri). The mixture was incubated at 56°C for 1 h, 100°C for 8 min, and then cooled to 40°C for 30 s. After vortexing for 30 s, samples were stored at -17°C. In cases where adult nematodes could not be found, individual infective juveniles were used.

PCR Conditions

The ITS-1 and ITS-2 regions of ribosomal DNA were amplified using 18S and 28S primers (no. 93, 5′ TTGAACCGGGTAAAAGTCG and no. 94, 5′TTAGTTTCTTTTCCTCCGCT) designed by Nadler et al. (2000). PCR amplification was performed in a reaction volume of 25 µl, containing 2.5 µl of 10x PCR-buffer, 14.6 µl of H2O, 3 µl of MgCl2 (25 mM), 0.5 µl of dNTPs (10 mM), 0.4 µl Taq polymerase (5 units/µl, Gene choice T-18), 1.25 µl of primer forward (10 µM), 1.25 µl of primer reverse (10 µM), and 1.5 µl of DNA. All PCR reactions were conducted in an Applied Biosystems 2720 thermocycler with a profile of: 1 cycle at 94°C for 2 min followed by 35 cycles of 94°C for 30 s, 60°C for 30 s and 72°C for 50 s. The last step was a post-amplification extension at 72°C for 4 min.

Table 1: Results of soil samplings for entomopathogenic nematodes (EPN) conducted in the Central Valleys of Oaxaca, from August to October, 2008

Locality	No. of samples	No. of positive samples and non-EPN	Positive samples for EPN (%)
Cuilapam de Guerrero	9	4	44.4
Santiaguito, Etla	2	1 (1)¥	50.0
Guadalupe, Ella	6	2(1)	33.3
Zaachila	4	1 (1)	25.0
Santa Cruz Xoxo	3	2 (1)	66.7
San Dionisio, Ocotlan	4	1	25.0
Lachigold, Tlacolula	1	1	100.0
Reyes Mantecon	8	0	0.0
Col. Emiliano Zapata	2	2	100.0
Sta. Inds Ayoquezco	3	2	66.7
Sn. Pedro Martir	4	4 (2)	100.0
Los Ocotes, Ejutla	4	4 (2)	100.0
San pablo Ella	5	3 (3)	60.0
Total	55	27 (11)	49.1

¥Number of non-entomopathogenic samples.

Table 2: Isolates with higher rates of mortality on *Galleria mellonella*

Sample number	Locality	Probable genera	Mortality (%)	Code
1	Cuilapam	Steinernema	97.5	CS1
2	Cuilapam	Steinernema	100	CS2
5	Cuilapam	Heterorhabditis	95	CH5
8	Cuilapam	Heterorhabditis	100	CH8
24	Lachiqolo Tlacolula	Heterorhabditis	100	LH24

Sequencing

Amplified products were purified using enzymatic treatment with exonuclease I and shrimp alkaline phosphatase (PCR product pre-sequencing kit, USB Corporation). Primers used for sequencing were the same as for PCR. Contiga were assembled using the program Codon Code Aligner (Version 3.6.1) and a basic local alignment search tool (BLAST) search performed on the final consensus sequence. Only matches above 99% identity were considered positive (Stock et al., 2001).

RESULTS

Soil Sampling

Table 1 shows the results of the soil samplings conducted in 13 communities of the Central Valleys of Oaxaca. From a total of 55 samples taken, 27 were positive for nematode infection, that is, 49.1% of the samples taken were infested by nematodes, but not necessarily entomopathogenic. According to the color of the dead larvae, at this point, 81% were infected by EPN of the genus *Steinernema* and 19% by *Heterorhabditis*.

Kaya and Stock (1997) mentioned that the *G. mellonella* larvae infected with EPN of the genus *Heterorhabditidae* may have reddish-orange coloring, while those infected with *Steinernema* spp. show gray-brown coloring. *G. mellonella* cadavers infected by different EPN isolates, showed distinctive colors, those infected by *H. bacteriophora* and *H. indica* were brick red, while those infected by *Steinernema feltiae* were brown in color (Nyamboli, 2008) (Table 2).

Experiment 1: Selection of Entomopathogenic Isolates

After isolation, the nematodes were bio assayed on *G. mellonella* larvae to detect the entomopathogenic ones. Eleven samples did not produce IJ's in adequate numbers or did not produce them at all, mostly because of fungal infection. Thus, the percentage of samples found positive for EPN was 29.1%.

About 1/3 of the isolates conformed a group where the control percentage was 95 to 100%. Four of the selected isolates were collected in the community of Cuilapam, Zaachila; these caused larval fatalities ranging from 95 to 100%. The other selected one was collected in the town of Lachigoloo, Tlacolula, which produced 100% mortality. The range of mortality observed for the remaining 10 isolates was 35 to 80%.

Table 3: Analysis of variance of transformed *Pyllophaga vetula* larvae mortality data

Source	Degrees of freedom	Sum of squares	Mean squares	F value	P>F
Model	244	28.35	1.18	43.11	<0.0001
Error	75	2.05	0.02		
Total	99	30.40			
Treatments	24	28.35	1.18	43.11	<0.0001
Isolates	4	0.13	0.03	14.35	0.0258
Dosages	4	32.32	8.13	485.12	<0.0001

R2 = 0.91; C. V. = 14.02%

Experiment 2: Lethal Dosages and Times for Selected Isolates

The analysis of variance of transformed mortality data of *P. vetula* larvae caused by five of the most effective isolates are shown in Table 3. The variance analysis showed that the general statistical model was adequate ($P < 0.05$); the R2 value was 0.91, indicating that the model explained 91% of variability of the experiment.

Treatments and dosages showed highly significant effects ($P < 0.01$), but isolate's effects were only significantly different ($P < 0.05$). Thus, there were statistical differences in at least one level of these factors. The comparison of means between treatments, and nematodes and dose factors are presented in Tables 4, 5 and 6, respectively. Table 4 shows that the main factor determining the statistical groups was the number of IJ's applied, as all the five isolates evaluated with 1000 nematodes per larva showed a high mortality. There was a clear statistical difference between the three higher dosages, but the two lower ones had no statistical differences between them (Table 5).

Table 6 shows that the nematodes from Heterorhabditidae, as compared with *Steinernema*tidae, are grouped in a different statistical group with a more effective grub control. Table 7 shows the lethal dosages and times for each of the five isolates applied for control of *P. vetula*. Both 50 and 95% dosages form two groups, one for the Heterorhabditidae, which require lower dosages to control white grub larvae, and another for the *Steinernema*tidae with higher dosages for control. The same groups are also valid for lethal times, where the *Steinernema*tidae showed the longer periods to control the larvae.

Determination of the Species of Nematodes

The isolates CH8, LH24 and CH5 were identified as *Heterorhabditis mexicana* Nguyen, Shapiro-Ilan, Stuart, McCoy, James and Adams; while isolate CS2 was identified as *Steinernema carpocapsae* Filipjev. The CS1 isolate was identified as *S. feltiae* Filipjev.

DISCUSSION

The percentage of positive samples for nematodes obtained in this study was 29.1%, which might seem high when compared with other studies, such as Stock et al. (1999), who conducted a sampling of different habitats in the state of California and found 26.3% of positive samples. In Oaxaca, Ruiz-Vega et al. (2003) found a maximum of 8.9% of positive samples in the region. The difference with the last study mentioned above may be because the sampling was random, including natural vegetation areas, while in the current study sampling was directed to maize fields where white grub infestation had occurred and thus the presence of EPN was more likely. This idea is supported by the study of Ruiz-Vega et al. (2003), who found that the highest percentages of nematode soil samples were found where the grubs were present.

In terms of percentage of positive samples by genera, this study found 68.7% of EPN in the genus *Steinernema* and 31.3% in the *Heterorhabditis*, which concurs with Stock et al. (1999), who reported 80% in *Steinernema* and 20% in *Heterorhabditis*; on the other hand, Ruiz-Vega et al. (2003) found 67% in *Steinernema* and 33% in *Heterorhabditis*.

The isolated nematodes were subject to an evaluation of relative pathogenicity. Those who were attacked by fungus or produced IJ's in small numbers were considered not entomopathogenic. It is known that when an insect is infected with the nematode's symbiotic bacteria, it is protected against the attack of other organisms. Rosa et al. (2000) consider that when larvae are infected by EPN, these will not rot, nor have foul odors. Also, each host will produce IJ's in large numbers. However, temperature and inoculums level have been found to affect the amount of IJ's produced.

In an evaluation with *H. heliothidis* and *S. glaseri*, temperature and inoculum level significantly affected time to first emergence, duration of emergence, and yield of juveniles from *G. mellonella* larvae (Zervos et al., 1991). Sepúlveda-Cano et al (2008) found similar results for inoculum level as they had a larger number of IJ's for 1000 than for 100 JI/25 µl; the amounts of IJ's produced per *Cosmopolites sordidus* larva was 99,386 for *H. bacteriophora* and 106,532 for *S. carpocapsae*. These numbers agree with Realpe-Aranda et al. (2007), who obtained an accumulated production of 86,250 IJ's/larva for *Steinernema colombiense* and 78,750IJ/larva for *H. bacteriophora*.

Table 4: Comparison of percent mortality means of *Phyllophaga vetula* between treatments

Treatment	Nematode species	Dosage (IJ's/larva)	Mortality (%)
25	H. mexicana	1000	100a¥
15	H. mexicana	1000	100a

10	S. carpocapsae	1000	97.5ab
20	H. mexicana	1000	97.5ab
5	S. feltiae	1000	95b
14	H. mexicana	500	90c
24	H. mexicana	500	90c
19	H. mexicana	500	87.5cd
18	H. mexicana	200	87.5cd
4	S. feltiae	500	47.5e
9	S. carpocapsae	500	47.5e
13	H. mexicana	200	37.5f
23	H. mexicana	200	37.5f
8	S. carpocapsae	200	32.5g
3	S. carpocapsae	100	30.0g
12	H. mexicana	100	22.5h
7	S. feltiae	200	17.5i
22	H. mexicana	100	15.0ij
21	H. mexicana	50	15.0ij
17	H. mexicana	100	10k
1	S. feltiae	50	7.5kl
16	H. mexicana	50	5.0l
2	S. feltiae	100	5.0l
11	H. mexicana	50	5.0l
6	S. carpocapsae	50	5.0l

¥Values with the same letter are not significantly different (Tukey P = 0.05).

Table 5: Comparison of mean mortality percentages of *Phyllophaga vetula* for different dosages of the EPN isolated

Dosage (IJ's/larva)	Mortality (%)
1000	98.0a¥
500	72.5b
200	42.5c
100	16.5d
50	7.5d

¥Values with the same letter are not significantly different (Tukey P = 0.05).

Table 6: Comparison of mean mortality percentage of *Phyllophaga vetula* for the different nematode species isolated

Isolate species	Mortality (%)
H. Mexicana H. Mexicana H. Mexicana S. carpocapsae S. feltiae	57.5a¥
H. mexicana	51.5a
H. mexicana	51.5a
S. carpocapsae	42.5b
S. feltiae	34.5c

¥Values with the same letter are not significantly different (Tukey P = 0.05).

Table 7: Dosages (D_{50} and D_{95}) and lethal times (LT_{50} and LT_{95}) for five nematode species isolated for control of *Phyllophaga vetula*

Isolate species	LD50	LD95	LT50	LT95
H. mexicana	326	726	3.1	7.1
H. mexicana	339	602	2.7	6.2
H. mexicana	348	613	2.8	6.3
S. carpocapsae	382	920	4.0	9.8
S. feltiae	405	998	4.1	10.3

According to the general statistical model, the more important factor to generate significance was the dose of nematodes applied, followed by isolates. In relation to application rates, the best dose was

1000 nematodes per larva, which coincides with Athanassiou et al. (2010), who reported that using *S. feltiae* in laboratory bioassays for the control of *Tribolium confusum* larvae Jacquelin du Val, the dose that caused the highest mortality (up to 56%) was 1000 nematodes per larva. Trdan, et al. (2009) found that for the control of larvae of Colorado potato beetle (*Leptinotarsa decemlineata),* the best dosages were 1000 and 2000 nematodes per larva.

The Heterorhaditidae were grouped in a different statistical group than the *Steinernema*tidae, the first being more effective for white grub control. The Heterorhabditidae had the lowest lethal doses; LD50 dosages varied from 326 to 348, while LD95 dosages ranged from 602 to 726 IJ's per larva. However, for the *Steinernema*tidae LD50 dosages varied from 382 to 405 and LD95 dosages varied from 920 to 998 IJ's per larva. Rodríguez et al. (2009) found an LD50 of 475 nematodes per larva for a native isolate of *Heterorhabditis* sp. applied to second and third stage larvae of *Phyllophaga elenans*, which is similar to the one found in the present study.

Also, the Heterorhabditidae presented the lowest LT50 values, ranging from 2.7 to 3.1 days, and the LT95 varied from 6.2 to 7.1 days; while for the *Steinernema*tidae the LT50 ranged from 4.0 to 4.1 and the LT95 had values of 9.8 to 10.3 days. Rodríguez et al. (2009) found LT50 values of 7.5 days for 625 IJ's of *Heterorhabditis* sp. applied to larvae of *P. elenans*. Ruiz-Vega et al. (2003) found an LT50 of three to seven days for *S. carpocapsae* and *S. feltiae* applied to *P. vetula* larvae.

CONCLUSIONS

This is the first report of *H. Mexicana* in the area, while *S. feltiae* and *S. carpocapsae* had been reported previously (Ruiz-Vega et al., 2003; Aquino et al., 2006). *S. feltiae* was located in a *Quercus spp.* forest, while *S. carpocapsae* was found in a plantation of *Agave angustifolia Haw.*

H. mexicana can be considered to show the highest percentage of control, but at 57.5%, this is still considered a moderate effectiveness. According to Georgis et al. (2006), a good control potential is above 70% effectiveness. Shapiro-Ilan et al. (2005) concluded that *H. mexicana,* relative to other entomopathogenic nematodes, generally possesses moderate abilities in bio-control traits. *H. mexicana* showed a medium potential for control of *P. vetula*, while *S. carpocapsae* and *S. feltiae* showed a low to very low potential.

ACKNOWLEDGMENTS

Tanks to Amanda Hodson, University of California at Davis, for carrying out the identification of the EPN by PCR techniques, and to Secretaría de Investigación y Posgrado, Instituto Politécnico Nacional, for funding this research project (SIP-IPN: 20082457).

REFERENCES

1. Aquino BT, Ruiz-Vega J, Iparraguirre CM (2006). Control biológico del picudo negro (*Scyphophorus interstitialis* Gyllenhal) con nemátodos y hongos entomopatógenos en agave en Oaxaca, México. Rev. UDO Agric. 6(1):92-101.
2. Athanassiou CG, Kavallieratos NG, Menti H, Karanastasi E (2010). Mortality of four stored product pests in stored wheat when exposed to dosages of three entomopathogenic nematodes. J. Econ. Entomol. 103(3):977-984.
3. Azorín F, Sánchez CJ (1986). Métodos y Aplicaciones del Muestreo. Alianza Universidad, Madrid, España.
4. Bravo M, E (2003). Gallina ciega: Diagnóstico de la situación que guarda en las regiones de la Mixteca y valles Centrales de Oaxaca.. *En* Aragón G. A., M. A. Morón y A. Marín J. (eds.) Estudios sobre coleópteros del suelo en América. Instituto de Ciencias, Benemérita Universidad Autónoma de Puebla. pp. 269-282.
5. Ellis RE, Sulston JE, Coulson AR (1986). The rDNA of C. *elegans*: Sequence and structure. Nucleic Acids Res. 14:2345–2364.
6. Georgis R, Koppenhöfer AM, Lacey LA, Bélair G, Duncan LW, Grewal PS, Samish M, Tan L, Torr P, Van TRWHM (2006). Successes and failures in the use of parasitic nematodes for pest control. Biol. Control 38:103-123.
7. Grewal PS, Selvan S, Gaugler R (1994). Thermal adaptation of entomopathogenic nematodes-niche breadth for infection, establishment, and reproduction. J. Therm. Biol. 19:245-253.

8. Griffin CT, Boemare NE, Lewis EE (2005). Biology and Behaviour. pp. 47-59 *in* Nematodes as Biological Control Agents, P.S. Grewal, R.U. Ehlers, and D. Shapiro-Ilan 1 ed. CABI Publishing, Wallingford.

9. Gruner DS, Ram K, Strong DR (2007). Soil mediates the interaction of coexisting entomopathogenic nematodes with an insect host. J. Invertebr. Pathol. 94:12-19.

10. Jackson TA (2003). Using entomopathogens in scarab pest management. *In* Estudios Sobre Coleópteros del suelo en América, G. A. Aragón, G. A., M. A. Morón y A. Marín J., 1a ed. Publicación especial de la Benemérita Universidad Autónoma de Puebla. Puebla, México. pp. 333-334.

11. Kaya HK, Gaugler R (1993). Entomopathogenics nematodes. Ann. Rev. Entomol. 38:181-206.

12. Kaya HK, Koppenhofer AM (1996). Effects of microbial and other antagonistic organism and competition on entomopathogenic nematodes. Biocontrol Sci. Technol. 6:357-371.

13. Kaya HK, Stock SP (1997). Techniques in insect nematology. pp. 281-324. *In* L. A. Lacey (ed.) Manual of techniques in insect pathology. Academic Press, New York.

14. Koppenhöfer A (2007). Nematodes. pp. 249-264 in: L. Lacey and H. Kaya, 2 ed. Field manual of techniques in invertebrate pathology: application and evaluation of pathogens for control of insects and other invertebrate pests. (2nd. ed.). Kluwer Academic Pub. Dordrecht, Netherlands.

15. Koppenhöfer AM, Kaya HK, Taormino SP (1995). Infectivity of entomopathogenic nematodes (Rhabditida,*Steinernema*tidae) at different soil depths and moistures. J. Invertebr. Pathol. 65:193-199.

16. Nadler SA, Adams BJ, Lyons ET, DeLong RL, Melin SR (2000). Molecular and morphometric evidence for separate species of *uncinaria* (nematoda: ancylostomatidae) in California sea lions and northern fur seals: hypothesis testing supplants verification. J. Parasitol. 86(5):1099-1106.

17. Nyamboli MA (2008). Mass production of entomopathogenic nematodesforplantprotection.2008.MasterofSciencedissertation, University of the Witwatersrand, Johannesburg. Available at http://

wiredspace.wits.ac.za/bitstream/handle/10539/5071/Mass%20 Production%20of%20Entomopathogenic%20Nematodes%20 for%20Plant%20Prot.pdf?sequence=1 Pp193.

18. Realpe-Aranda FJ, Bustillo AE, Pardey, López-Nuñez JC (2007). Optimización de la cría de Galleria mellonella (L.) para la producción de nematodos entomopatógenos parásitos de la broca del café. Cenicafé (Colombia) 58(2):142-157.
19. Rodríguez D, Torres M, Uribe L, Flores L (2009). Susceptibilidad de los estadios L2 y L3 de *Phyllophaga elenans* a una cepa nativa de *Heterorhabditis* sp. en condiciones de invernadero. Agron. Costarricense 33(2):171-182.
20. Rosa J, Bonfassi E, Amaral J, Lacey LA, Simoes N, Laumond C (2000). Natural occurrence of entomopathogenic nematodes (Rhabditida: *Steinernema*, Heterorhabditis) in the Azores. J. Nematol. 32:215-222.
21. Ruiz-Vega J, Aquino-Bolaños T, Kaya HK, Stock P (2003). Colecta y evaluación de nematodos entomopatógenos para el control de gallinas ciegas *Phyllophaga vetula* (Horn) en Oaxaca, México. Folia Entomol. Mex. 42(2):169-175.
22. SAS (2002). SAS Institute Inc., SAS/STAT. User's Guide, Version 9.0, Cary NC. Statistical Analysis System Institute Inc., North Caroline.
23. Sepúlveda-Cano PJ, López-Núñez C, Soto-Giraldo A (2008). Efecto de dos nematodos entomopatógenos sobre *Cosmopolites sordidus* (Coleoptera:Dryophthoridae) Rev. Colombiana Entomol. (34):62-67.
24. Shapiro-Ilan DI, Stuart RJ, McCoy CW (2005). Characterization of biological control traits in the entomopathogenic nematode *Heterorhabditis mexicana* (MX4 strain) Biological Control : Theor. Appl. Pest Manag. 32(1):97-103.
25. Stock SP, Pryor BM, Kaya HK (1999). Distribution of entomopathogenic nematodes (*Steinernema*tidae and Heterorhabditidae) in natural habitats in California, USA. Biodivers. Conserv. 8:535-549.
26. Stock, SP, Campbell JF, Nadler SA (2001). Phylogeny of *Steinernema* Travassos, 1927 (Cephalobina: *Steinernema*tidae) Inferred From Ribosomal DNA Sequences and Morphological Characters. J. Parasitol. 87(4):877-889.

27. Trdan S, Vidrih M, Andjus L, Laznik Z (2009). Activity of four entomopathogenic nematode species against different developmental stages of Colorado potatobeetle, *Leptinotarsa decemlineata* (Coleoptera: Chrysomelidae). Helminthology 46(1):14-20.

28. Villalobos JF (1992). The potential of entomopathogens for the control of white grub pests of corn in México. pp. 253-260. *In:* Jackson T. A. y Glare, T. R. (eds.) Use of pathogens in scarab pest management. Intercept Ltd., Andover, Hampshire. 295 pp.

29. Wang X, Grewal PS (2002). Rapid genetic deterioration of environmental tolerance and reproductive potential of an entomopathogenic nematode during laboratory maintenance. Biol. Control 23:71-78.

30. White GF (1927). A method for obtaining infective nematode larvae from cultures. Science 66:302-303.

31. Zervos S, Johnson SC, Webster JM (1991). Effect of temperature and inoculum size on reproduction and development of *Heterorhabditis heliothidis* and *Steinernema glaseri* (Nematoda: Rhabditoidea) in *Galleria mellonella* . Can. J. Zool. 69(5):1261-1264.

32. Zhou W, Rousset F, O'Neil S (1998). Phylogeny and PCR-based classification of *Wolbachia* strains using *wsp* gene sequences. P. Roy. Soc. Lond. B Bio. 265:509-515.

Chapter 9

Mermithid Nematodes Found in Adult Anopheles Southeastern Senegal

Kevin C Kobylinski, Massamba Sylla, William Black IV, and Brian D Foy

Arthropod-borne Infectious Disease Laboratory, Department of Microbiology, Immunology and Pathology, Colorado State University, 1692 Campus Delivery, Fort Collins, CO 80523-1692, USA

ABSTRACT

Background

Over two dozen mermithid nematodes have been described parasitizing mosquitoes worldwide, however, only two species were found in Africa. Mermithid nematodes kill their mosquito host upon emergence, which suggests that they could be developed as biological control agents of mosquitoes. Both Romanomermis culicivorax and Romanomermis iyengari have been reared for mass release to control numerous Anopheles species vector populations, and in one instance this may have led to reduced malaria prevalence in a human population.

Methods

Anopheles mosquitoes were collected during a malaria study in southeastern Senegal. Two different adult blood fed mosquitoes had a single mermithid nematode emerge from their anus while they were being held post-capture. Primers from the 18 S rDNA were developed to sequence nematode DNA and screen mosquitoes for mermithid DNA. 18 S rDNA from the Senegalese mermithid and other mermithid entries in GenBank were used to create a Maximum Parsimony tree of the Mermithidae family.

Results

The mermithid was present in 1.8% (10/551) of the sampled adult Anopheles species in our study area. The mermithid was found in An. gambiae s.s., An. funestus, and An. rufipes from the villages of Ndebou, Boundoucondi, and Damboucoye. Maximum parsimony analysis confirmed that the nematode parasites found in Anopheles were indeed mermithid parasites, and of the mermithid sequences available in GenBank, they are most closely related to Strelkovimermis spiculatus.

Conclusions

To our knowledge, this is the first report of mermithids from adult Anopheles mosquitoes in Senegal. The mermithid appears to infect Anopheles mosquitoes that develop in diverse larval habitats. Although maximum parsimony analysis determined the mermithid was closely related to Strelkovimermis spiculatus, several characteristics of the mermithid were more similar to the Empidomermis genus. Future mermithid isolations will hopefully allow: formal taxonomic identification, laboratory colonization, determination of life history traits and species specificity, and characterize its usefulness as a biological control agent

BACKGROUND

Nematodes from the family Mermithidae (Enoplea: Mermithida) parasitize terrestrial and aquatic invertebrates, including many insect disease vector species [1]. Many insect disease vectors are parasitized by mermithid nematodes during their aquatic larval stage, including blackflies and mosquitoes [2]. Seven genera of Mer mithidae are parasites of mosquitoes: Culicimermis, Empi domermis, Hydromermis, Octomyomermis, Perutilimermis, Romanomermis, and Strelkovimermis [3], and over 25 species have been described [4]. While numerous mermithids have been found in mosquitoes on every continent except Antarctica [4-8], only Empidomermis cozii and Reesimermis (Octomyomermis) muspratti, have been described from mosquitoes in Africa [9-11]. Octomyomermis muspratti was isolated from various Aedes and Culex spp. larvae from tree holes in Livingstone, Zambia [11]. Subsequent colonies were established and O. muspratti was found to be capable of parasitizing Aedes aegypti, Aedes polynesiensis Collins, CO 80523-1692, USA © 2012 Kobylinski et al.; licensee BioMed Central Ltd. This is an Open Access article distributed under the terms of the Creative Commons Attribution License (http://creativecommons.org/licenses/by/2.0), which permits unrestricted use, distribution, and reproduction in any medium, provided the original work is properly albimanus [12]. Empidomermis cozii was found in adult Anopheles funestus from Burkina Faso [9,10]. Regardless of genera, aquatic mermithids that infect mosquitoes have similar life cycles. Eggs hatch in either standing or flowing water and the pre-parasitic forms attach to and bore into mosquito larvae. Parasitism is typically most successful when the pre-parasitic form invades a first or second instar mosquito. Pre-parasitic mermithids develop in the mosquito host, molt once, and emerge from the mosquito as post-parasitic forms. Mermithid emergence is typically lethal for the mosquito and may happen in the mosquito larval or adult stage. In the aquatic habitat the post-parasitic mermithids molt into adults, mate and lay eggs. The development time of each mermithid life stage varies among species, and this dictates at which mosquito life stage the post-parasitic form emerges [13]. Mosquito death upon mermithid emergence has sparked interest in using mermithids as biological control agents. From a classical biological control standpoint, Juliano (2007) cited several characteristics of mermithids that make them potentially successful biological control agents of mosquitoes: they are specialists, they develop in synchrony with mosquitoes, their populations can increase rapidly or be augmented by human reinundation, and they require only one victim/host to complete their life cycle [14]. Platzer (2007) summarizes multiple characteristics of mermithids that make them attractive for biological control of mosquitoes, including ease of application, environmental safety, host specificity, laboratory manipulation of life history, lethality, mass rearing in vivo, and potential for long-term recycling in the environment. Numerous investigators have developed mass rearing techniques for mermithids [15-17] that can support large scale field releases. Mermithids have been applied to aquatic habitats as either pre-parasitic forms with a hand operated spray pump [17] or aerial spray pump [18] and post-parasitic forms applied to moist soil prior to flooding [19,20]. Major field releases of both Romanomermis culicivorax and Romanomermis iyengari have occurred for Anopheles population control. Romanomermis iyengari has been used to suppress malaria vector populations of An. albimanus in Cuba [21], Anopheles pseudopunctipennis in Oaxaca State, Mexico [22], and Anopheles albitarsis and Anopheles rondoni in Roraima State, Brazil [23]. Romanomermis culicivorax has been used to suppress malaria vector populations of Anopheles freeborni in California, USA [18,19], An. albimanus and An. pseudopunctipennis in El Salvador [16,24], An. albimanus in Mella Isla de la Juventud, Cuba [25], and An. albimanus in El Valle, Colombia [26]. The publication by Rojas et al. (1987) is notable because release of R. culicivorax may have both suppressed the vector population and reduced malaria prevalence in local children for up to two years post mermithid release. The successful control of An. albimanus by R. culicivorax, a mermithid that exits the mosquito in the larval stage, was facilitated by the fact that An. albimanus larvae occur in large standing bodies of water that can be readily identified and treated with mermithids [26]. This evidence demonstrates that mermithid nematodes can be successfully used as biological control agents of mosquitoes. This report documents numerous mermithid nematodes that were observed in adult female Anopheles spp. collected during a malaria study conducted in southeastern Senegal [27]. Sequencing and phylogeny analysis were used to place the nematodes in the Mermithidae family and compare their relatedness to other mermithid species.

METHODS

All mosquitoes were collected in 2009 from the villages of Ndebou, Boundoucondi, Nathia, and Damboucoye in southeastern Senegal [28]. Blood fed mosquitoes were collected from huts during the mornings by backpack aspiration and held in a field insectary for up to five days [27]. Other mosquitoes were collected from CDC light traps hung next to bed nets; these mosquitoes were processed on the same day of capture. Mosquitoes were identified morphologically [29] and molecularly [30,31] to species. During routine morphological examination, mosquitoes were also inspected visually for possible mermithid infection. Adult mermithids range in length from 10 to 100 mm [1] so they can be readily observed without dissection of the mosquito (see Figure 1a). Mermithid nematodes were only observed in mosquito abdomens. Abdomens were separated from the thorax of the adult mosquitoes and stored on silica gel desiccant. Two mermithid-positive An. gambiae s.s. were dissected and the nematodes were separated from the mosquitoes so that mermithid DNA could be extracted with the Qiagen DNeasy kit (Qiagen Sciences, Maryland, USA) and a Qiacube robot (Qiagen Sciences, Maryland, USA). Primers were designed to amplify regions of the 18 S ribosomal

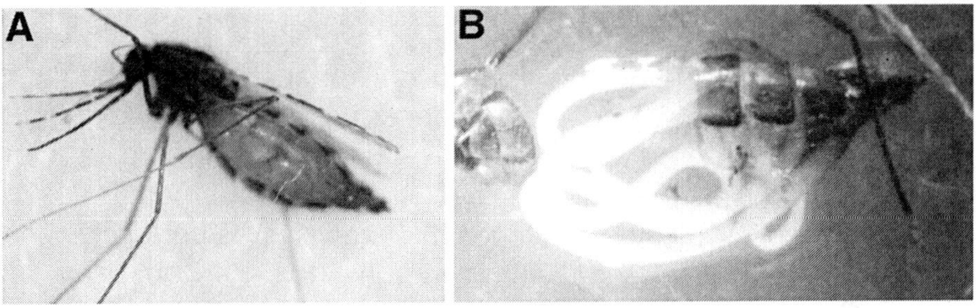

Figure 1: Photograph of a mermithid-parasitized An. rufipes (A) and the dissected mosquito abdomen with mermithid parasite (B).

RNA gene by the polymerase chain reaction (PCR). Primers were designed by alignment of sequences from other mermithids obtained from GenBank and making consensus primers. Primer sequences were: Merm forward – 5′-CAAGGACGAAAGTTAGAGGTTC-3′ and Merm reverse – 5′-GGAAACCTTGTTACGACTTTTA-3′. Twenty μl of a PCR enzyme mixture was used to amplify DNA from each sample; this mixture consisted of: 10 μl Fermentas 2X Master Mix, 2 μl forward and reverse primers at 10 μM, 2 μl DNA, and 4 μl of RNAse-free water. The thermocycler protocol was 5 minutes at 95°C, 30 seconds denaturing at 95°C, 30 seconds annealing at 50°C, and 45 seconds extension at Kobylinski et al. Parasites & Vectors 2012, 5:131 http://www.parasitesandvectors.com/content/5/1/131 72°C for 40 cycles with an additional 10 minute extension step, all performed on a PTC-200 Peltier Thermal Cycler (MJ Research, Waltham, MA, USA). This protocol was used to assess the presence of mermithid DNA from the abdomens of blood fed mosquitoes that were processed for blood meal analysis [27]. The primers amplified an 804 base pair fragment of mermithid DNA and also a 900 base pair fragment of Anopheles spp. DNA. Thus, when these primers were used to screen wild Anopheles mosquitoes one band would always be present when a mosquito was processed, but when the mosquito was infected with a mermithid parasite, there would be two bands present (see Figure 2). 18 S rDNA amplicons from mermithid-only isolates were cloned in a TOPO-TA plasmid (Invitrogen, Carlsbad, CA, USA), plasmid DNA was harvested from transformed E. coli with a QIAprep Spin Miniprep Kit (Qiagen,

Gaithersburg, Maryland, USA) and DNA was submitted for sequencing by the Proteomics and Metabolomics Facility at Colorado State University, Fort Collins, Colorado. Sequenced mermithid DNA was then used to identify 147 of the most similar sequences published in GenBank, based on E-score. Sequences were provisionally aligned using Clustal W [32]. This alignment was further adjusted by hand, based on the current understanding of RNA secondary models [33,34]. Maximum parsimony (MP) phylogenetic analysis was performed using the Phylogenetic Analysis Using Parsimony (PAUP) 4.0b10 package [35].

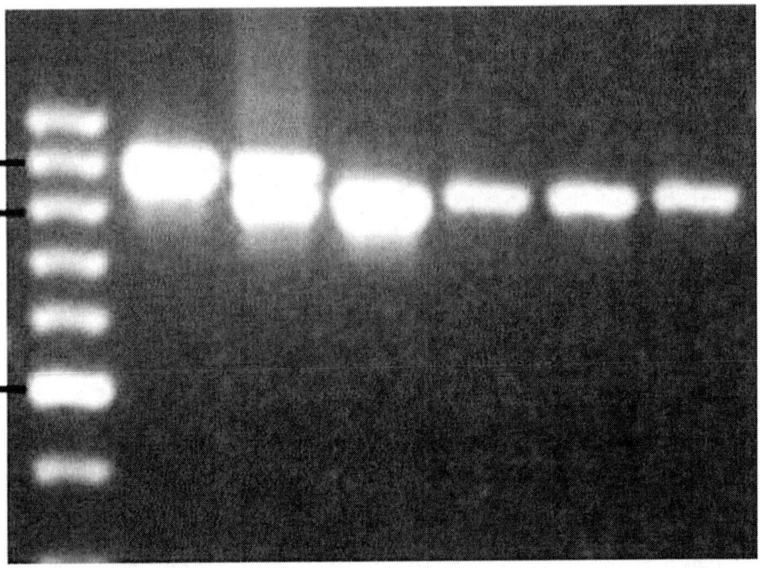

Figure 2: Gel electrophoresis image of multiple samples amplified with mermithid primers: Lane 1 – 100 base pair ladder, Lane 2 – uninfected laboratory-reared An. gambiae s.s., Lane 3 – Fieldcollected An. gambiae s.s. infected with mermithid parasite, Lanes 4 – 7 – dissected mermithid samples from field-collected mosquitoes (free of Anopheles spp. tissue).

MP was chosen because it is the only phylogenetic method that uses gaps as phylogenetically informative characters in sequence alignments and gaps were common in the 18 S ribosomal RNA alignments. Four levels of outgrouping with 100 rounds of bootstrapping were performed. Nematodes from the Mermithidae family formed a monophyletic clade and therefore only sequences from these were compared for the final tree [GenBank Accession Nos.: Agamermis changshaensis – DQ628908.1, Agamermis sp. BH-2006 - DQ665653.1, Agamermis xianyangensis – EF617352.1, Allomermis solenopsii – DQ533953.1, Amphimermis sp. A-2007 – EF617354.1, Amphimermis sp. B-2007 – EF617355.1, Cryptonchus tristis – EF207244.1, Dintheria tenuissima – AY593942.1, Gastromermis sp. – AY146543.1, Gastromermis sp. BH-2006 – DQ533954.1, Heleidomermis sp. BH-2006 – DQ533955.1, Hexamermis agrotis – DQ530 350.1, Isomermis lairdi – FN400900.1, Mermis nigrescens – AF036641.1, Mermithid sp. Jh-

2004 – AY284743.1, Mermithidae sp. A-AV-2003 – AY374415.1, Mermithidae sp. C-AV-2003 – AY374417.1, Mermithidae sp. MHMH-2008 – FJ040480.1, Octomyomermis huazangensis – EF617353.1, Ovomermis sinensis – DQ520879.1, Romanomermis culicivorax – DQ418791.1, Romanomermis sichuanensis – EF612769.1, Romanomermis wuchangensis – DQ520878.1, Strelkovimermis spiculatus – DQ665654.1, Thalassoalaimus pirum – FJ040500.1, and Thaumamermis cosgrovei – DQ665655.1].

RESULTS

Mermithid DNA was amplified in a total of ten of 551 (1.8%) Anopheles spp. abdomens collected from Figure 1 Photograph of a mermithid-parasitized An. rufipes (A) and the dissected mosquito abdomen with mermithid parasite (B). Figure 2 Gel electrophoresis image of multiple samples amplified with mermithid primers: Lane 1 – 100 base pair ladder, Lane 2 – uninfected laboratory-reared An. gambiae s.s., Lane 3 – Fieldcollected An. gambiae s.s. infected with mermithid parasite, Lanes 4 – 7 – dissected mermithid samples from field-collected mosquitoes (free of Anopheles spp. tissue). Kobylinski et al. Parasites & Vectors 2012, 5:131 Page 3 of 7 http://www.parasitesandvectors.com/content/5/1/131 southeastern Senegal. Mermithids were visually observed in four of these captured Anopheles spp. mosquitoes, and six others were detected only by PCR. Two of the four mermithid nematodes observed visually had emerged from the anus of aspirated mosquitoes that died during the five day holding period, similar to the description and photograph of E. cozii in Poinar (1977). One An. rufipes that had survived to five days post capture contained a live mermithid (Figure 1), but attempts to colonize the mermithid failed. Mermithid-positive Anopheles spp. were found from 3 of the 4 villages from which we sampled mosquitoes: Ndebou (5), Boundoucondi (3), and Damboucoye (2). Five of 74 (6.8%) An. gambiae s.s. tested from Ndebou, two of 91 (2.2%) An. funestus and one An. rufipes tested from Boundoucondi, and two of 51 (3.9%) An. funestus tested from Damboucoye were positive for mermithid parasites or mermithid DNA. Maximum parsimony analysis confirmed that the nematode parasites found in Anopheles from Senegal were indeed mermithid parasites. A clade containing Senegalese mermithids and Strelkovimermis spiculatus had 72% bootstrap support (Figure 3).

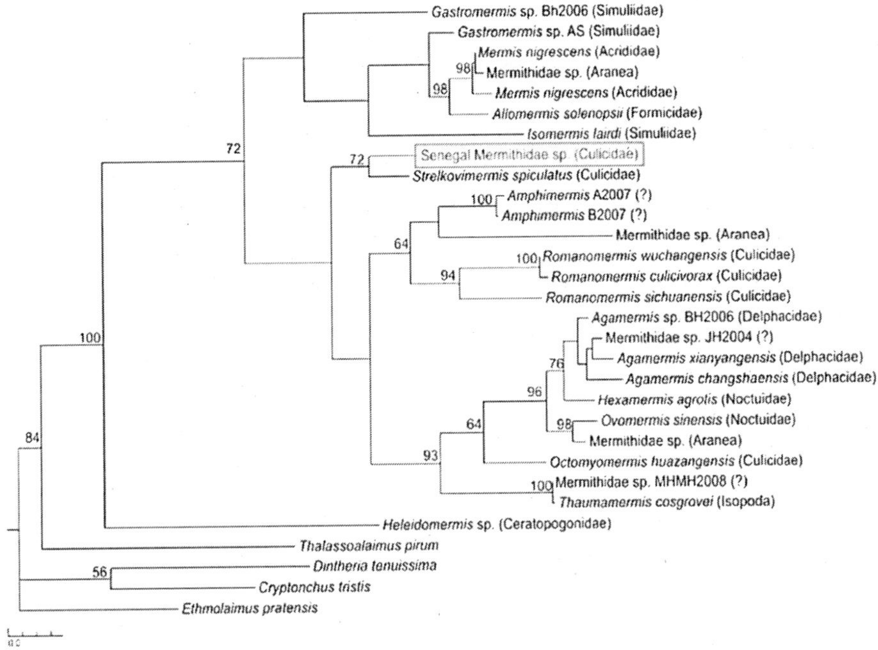

Figure 3: The Maximum Parsimony tree demonstrates that the nematode found in various Anopheles spp. from Senegal is indeed a mermithid parasite. Numbers at nodes represent the degree of bootstrap support. The scale represents the number of steps. Mermithid species names in italics are followed by the family or order the parasite was isolated from in parentheses.

DISCUSSION

To our knowledge, this is the first report of a mermithid nematode from Anopheles mosquitoes in Senegal. The MP bootstrap analysis placed the Senegalese mermithid in the same clade with Strelkovimermis spiculatus. There are no Empidomermis, Culicimermis, Hydromermis, or Perutilimermis 18 S rDNA entries in GenBank, therefore the exact placement of the Senegalese mermithid within the family Mermithidae is unknown. The two previous mermithid species isolated from Africa, E. cozii [9,10] and O. muspratti [11,12] do not have published 18 S ribosomal RNA sequences. The distance of the Senegalese mermithid from Octomyomermis huazangensis in the MP tree (Figure 3) suggests that this parasite does not belong to the Octomyomermis genera. It is possible that the Senegalese mermithids were re-isolations of E. cozii or a different mermithid species not listed in GenBank and therefore not a novel mermithid parasite isolation. The Merm primers developed here could be useful for screening mosquito populations for novel mermithid isolates. Re-isolation of the Senegalese mermithid and proper preparation of specimens for taxonomic evaluation will be necessary to make a more accurate species identification. Only two Strelkovimermis species have been identified as parasites of mosquitoes, Strelkovimermis spiculatus [8] and Strelkovimermis peterseni [36]. Both S. spiculatus [37] and S. peterseni [36] have only been reported to exit the mosquito in the larval stage. Strelkovimermis spiculatus parasitize Anopheles species in laboratory settings but not as efficiently as Culex and Aedes species [37,38]. Conversely, S. peterseni was only capable of parasitizing Anopheles species larvae and not any of the Culex, Aedes, or Psorophora larvae tested [39]. Similar to other mermithid species, non-target aquatic arthropods Figure 3 The Maximum Parsimony tree demonstrates that the nematode found in various Anopheles spp. from Senegal is indeed a mermithid parasite. Numbers at nodes represent the degree of bootstrap support. The scale represents the number of steps. Mermithid species were rarely parasitized by S. spiculatus, and therefore Strelkovimermis species may be desirable for use as a biological control agent [37,38]. To date, no field release of either S. spiculatus or S. peterseni has occurred [13]. The Senegalese mermithid may have life cycle characteristics similar to Empidomermis spp. Three isolations of Empidomermis spp. have been documented: E. cozii from Burkina Faso [9,10], Empidomermis riouxi from France [40], and Empidomermis spp. from New York, USA [41]. Empidomermis cozii was reported to emerge from the anus of adult mosquitoes [10], similar to what we observed with the Senegalese mermithid. Empidomermis cozii cannot fully develop in the adult mosquito unless a blood meal has been ingested by the mosquito [10], and all of the mermithid-positive mosquitoes during our study were blood-fed at the point of capture. Senegalese mermithids were observed or detected in roughly 2% of the Anopheles in our study area which demonstrates that mermithids are naturally present in Anopheles populations in West Africa. The Merm primers amplified both mosquito and nematode DNA, therefore sequence data was only available for mermithids that had emerged from the adult mosquito and it is not possible to tell if all mermithids detected were the same species. Mermithid parasites were found in

An. gambiae s.s., An. funestus, and An. rufipes. If these mermithid specimens were the same nematode species in all three mosquito species, this would indicate that the mermithid has a broad host range within the Anopheles genera. In the future, additional sequencing will be necessary to determine if the Senegalese mermithids from all infected Anopheles species are the same mermithid species or possibly distinct molecular variants as was observed with Mesomermis flumenalis in multiple Simulium species [42]. Anopheles gambiae s.s., An. funestus, and An. rufipes inhabit a diverse and disparate range of larval habitats with typically little overlap [43], therefore the transmission of the Senegalese mermithid may not be restricted by larval habitat type. A mermithid with a broad mosquito host range is more likely to be recycled in the environment which would reduce the number of releases necessary for biological control of mosquito populations [13]. Several characteristics of the development of Empido mermis and the Senegalese mermithid described here may make it an ideal biological control agent for the African malaria vector Anopheles gambiae. Larval control of Anopheles gambiae is difficult because its larval habitats are extremely abundant, ephemeral or inaccessible [44]. A mermithid parasite that emerges from the adult could be dispersed to cryptic An. gambiae larval habitats. Biological control of mosquito larvae in ephemeral larval habitats is difficult due to issues of larval density dependence, compensatory mortality, and overcompensation [14]; however, a mermithid that emerges from the adult mosquito would bypass these issues of biological control. Furthermore, mermithid parasites that emerge from the adult mosquito may be more effective for vector-borne disease control than a mermithid that emerges from a larval mosquito as this is likely to reduce the daily probability of adult survival, the most critical variable in the vectorial capacity equation [45]. It was speculated that Isomermis lairdi, a mermithid parasite of Simulium damnosum that emerges in the adult life stage of the black fly, may limit Onchocerca volvulus transmission by killing the adult black fly vector before O. volvulus could fully develop and be transmitted [46]. All three Empidomermis species were reported to sterilize adult mosquitoes [9,40,41]. Empidomermisparasitized females had arrested ovarian development [40,41] and inhibited mating activity of parasitized males [40]. These characteristics would maximize transmission of the mermithid to other mosquito offspring as opposed to infection of the parasitized females own offspring. For ideal mermithid-based control of malaria parasite transmission, the mermithid would need to emerge from the adult mosquito prior to Plasmodium completion of the extrinsic incubation period in the Anopheles vector. There is also the possibility that the mermithid may inhibit Plasmodium development in the mosquito, which could further impact vector-borne disease control. While a mermithid that emerges from the mosquito adult stage may be more difficult to rear in large numbers for mass release, this same life history trait may be used to facilitate a novel, rapid dispersal method of the mermithid into the environment if mermithid-infected adult mosquitoes were released. The re-isolation of the Senegalese mermithid from adult mosquitoes should be feasible. Specimens should be prepared and submitted to a nematode taxonomist to make a formal identification and description of the species. Laboratory colonies should be established that utilize rearing techniques published for several other mermithids from mosquitoes [15,17]. Sampling of wild adult and larval mosquito populations irrespective of mosquito genera over a longer duration of time combined with DNA barcoding of mermithids will determine critical information on host and temporal range of the mermithid. The life cycle and interactions of the Senegalese mermithid with various Anopheles species should be determined to assess the potential effect that mermithid parasitism may have on vector populations and malaria parasite transmission in order to define the utility of the mermithid as a biological control agent.

CONCLUSIONS

To our knowledge, this is the first report of mermithids from adult Anopheles mosquitoes in Senegal. Mermithids were present in 1.8% of the sampled adult Anopheles Kobylinski et al. Parasites & Vectors 2012, 5:131 Page 5 of 7 http://www.parasitesandvectors.com/content/5/1/131 species in our study area. The mermithid was found in An. gambiae s.s., An. funestus, and An. rufipes which suggests parasite infection occurred in numerous, diverse mosquito larval habitats. Although maximum parsimony analysis determined the mermithid was closely related to S. spiculatus, several characteristics of the Senegalese mermithid were more similar to those more commonly associated with the Empidomermis genus, such as emergence from the adult mosquito. Future mermithid isolations of the Senegalese mermithid will hopefully allow for formal taxonomic identification, laboratory colonization, determination of life history traits and species specificity, and characterize its usefulness as a biological control agent.

AUTHORS' CONTRIBUTIONS

KCK collected and processed specimens, analyzed data, and wrote the manuscript. MS collected specimens and wrote the manuscript. WBIV performed the phylogenetic analyses and wrote the manuscript. BDF collected specimens, provided reagents, and wrote the manuscript. All authors read and approved the final version of the manuscript.

ACKNOWLEDGEMENTS

We would like to thank the people of Ndebou, Boundoucondi, Nathia, and Damboucoye for allowing us to collect mosquitoes from their villages, Meg Gray for processing mosquito samples, and George Poinar Jr. for helpful discussion. This study was funded by grants R21AI079528 and R01AI094349 from the US National Institutes of Health. Received: 16 May 2012 Accepted: 18 June 2012 Published: 28 June 2012

REFERENCES

1. Poinar GO: Nematodes for biological control of insects. Boca Raton, FL: CRC Press; 1979.
2. Petersen JJ: Nematodes as biological control agents: Part I. Mermithidae. Adv Parasit 1985, 24:307–346.
3. Poinar GO: Nematoda and Nematomorpha. In Ecology and Classification of North American Freshwater Invertebrates. 2nd edition. Edited by Thorp JH, Covich AP. San Diego, CA: Academic; 2001:255–295.
4. Blackmore MS: Mermithid parasitism of adult mosquitoes in Sweden. Am Midl Nat 1994, 132:192–198.

5. Kalucy EC: Parasitism of Anopheles annulipes Walker by a mermithid nematode. Mosquito News 1972, 32:582–585.
6. Blackmore MS, Berry RL, Foster WA, Walker ED, Wilmot TR, Craig GB: Records of mosquito-parasitic mermithid nematodes in the North-central United States. J Am Mosq Control Assoc 1993, 9:338–343.
7. Vythilingam I, Sidavong B, Seng TC, Phonemixay T, Phompida S, Krishnasamy M: First report of mermithid parasitism (Nematoda: Mermithidae) in mosquitoes (Diptera: Culicidae) from Lao PDR. Tropical Biomedicine 2005, 22:77–79.
8. Poinar GO, Camino NB: Strelkovimermis spiculatus n. sp. (Mermithidae: Nematoda) parasitizing Aedes albifasciatus Mac (Culicidae: Diptera) in Argentina. J Nematol 1986, 18:317–319.
9. Coz J: Contribution a l'etude du parasitisme des adultes d'Anopheles funestus par Gastromermis sp. (Mermithidae). Bull Soc Pathol Exot 1966, 59:881–889.
10. Poinar GO: Empidomermis cozii n. gen., n. sp. (Mermithidae: Nematoda) a parasite of Anopheles funestus (Culicidae: Diptera) in West Africa. Canadian J Zool 1977, 55:1475–1479.
11. Muspratt J: Observation on the larvae of tree-hole breeding Culicini (Diptera: Culicidae) and two of their parasites. J Entomol Soc South Africa 1945, 8:13–20.
12. Obiamiwe BA, MacDonald WW: A new parasite of mosquitoes,Reesimermis muspratti sp. nov. (Nematoda: Mermithidae), with notes onIts life-cycle. Ann Trop Med Parasitol 1973, 67:439–444.
13. Platzer EG: Mermithid nematodes. J Am Mosq Control Assoc 2007,23:58–64.
14. Juliano SA: Population dynamics. J Am Mosq Control Assoc 2007, 23:265–275.
15. Petersen JJ: Biology of Octomyomermis muspratti, a parasite of mosquitoes as it relates to mass production. J Invertebr Pathol 1977,30:155–159.
16. Petersen JJ, Chapman HC, Willis OR, Fukuda T: Release ofRomanomermis culicivorax for control of Anopheles albimanus in ElSalvador. II. Application of nematode. AmJTrop Med Hyg 1978,27:1268–1273.

17. Petersen JJ and Willis OR: Procedures for mass rearing of a mermithid parasite of mosquitoes. Mosquito News 1972, 32:226–230.

18. Levy R, Hertlein BC, Petersen JJ, Doggett DW, Miller TW: Aerial application of Romanomermis culicivorax (Mermithidae: Nematoda) to control Anopheles and Culex mosuqitoes in southwest Florida. Mosquito News 1979, 39:20–25.

19. Brown-Westerdahl B, Washino RK, Platzer EG: Successful establishment and subsequent recycling of Romanomermis culicivorax (Mermithidae:Nematoda) in a California rice field following post-parasite application. J Med Entomol 1982, 19:34–41.

20. Kerwin JL, Washino RK: Recycling of Romanomermis culicivorax (Mermithidae: Nematoda) in rice fields in California, USA. J Med Entomol 1985, 22:637–643.

21. Santamarina Mijares A: Actividad parasitaria de Romanomermis iyengari (Nematoda: Mermithidae) en criaderos naturales de larvas de mosquito. Miscellania Zoolaogica 1993, 17:59–65.

22. Santamarina Mijares A, Perez Pacheco R, Tomas Martinez SH, Canton LE, Ambrosio GF: The Romanomermis iyengari parasite for Anopheles pseudopunctipennis suppression in natural habitats in Oaxaca State, Mexico. Revista Panamericana de Salud Publica 1999, 5:23–28.

23. Santamarina Mijares A, Bellini AC: Mass production of Romanomermis iyengari (Nematoda: Mermithidae) applied to anopheline breeding sites in Boa Vista (Roraima), Brazil. Revista Panamericana de Salud Publica 2000, 7:155–161.

24. Willis OR, Chapman HC, Petersen JJ: Additional field testing of the mermithid parasite Romanomermis culicivorax against Anopheles albimanus in El Salvador. Mosquito News 1980, 40:71–73.

25. Santamarina Mijares A, Perez Pacheco R: Reduction of mosquito larval densities in natural sites after introduction of Romanomermis culicivorax (Nematoda: Mermithidae) in Cuba. J Med Entomol 1997, 34:1–4.

26. Rojas W, Northup J, Gallo O, Montoya AE, Montoya F, Restrepo M, Nimnich G, Arango M, Echavarria M: Reduction of malaria prevalence after introduction of Romanomermis culicivorax

(Mermithidae: Nematoda) in larval Anopheles habitats in Colombia. Bulletin of the W Health Org 1987, 65:331–337.

27. Sylla M, Kobylinski KC, Gray M, Chapman PL, Sarr MD, Rasgon JL, Foy BD: Mass drug administration of ivermectin in south-eastern Senegal reduces the survivorship of wild-caught, blood fed malaria vectors.
28. Malar J 2010, 9:e365.
29. Kobylinski KC, Sylla M, Chapman PL, Sarr MD, and Foy BD: Short Report: Ivermectin
30. Mass Drug Administration to Humans Disrupts Malaria Parasite Transmission in Senegalese Villages. AmJTrop Med Hyg 2011, 85:3–5.
31. Diagne N, Fontenille D, Konate L, Faye O, Lamizana MT, Legros F, Molez JF, .and.Trape JF: The Anopheles of Senegal - an annotated and illustrated check list. Bulletin De La Societe De Pathologie Exotique 1994, 87:267–277.
32. Scott JA, Brogdon WG, and Collins FH: Identification of single specimens of the Anopheles gambiae complex by the polymerase chain reaction. AmJTrop Med Hyg 1993, 49:520–529.
33. Koekemoer LL, Kamau L, Hunt RH, Coetzee M: Cocktail polymerase chain reaction assay to identify members of the Anopheles funestus (Diptera: Culicidae) group. AmJTrop Med Hyg 2002, 66:804–811.
34. Thompson JD, Higgins DG, Gibson TJ, Clustal W: Improving the sensitivity of progressive multiple sequence alignment through sequence weighting, position-specific gap penalties and weight matrix choice. Nucleic Acids Res 1994, 22:4673–4680.
35. Gutell RR, Lee JC, Cannone JJ: The accuracy of ribosomal RNA comparative structure models. Curr Opin Struct Biol 2002, 12:301–310.
36. Gillespie JJ, McKenna CH, Yoder MJ, Gutell RR, Johnston JS, Kathirithamby J, Cognato AI: Assessing the odd secondary structural properties of nuclear Kobylinski et al. Parasites & Vectors 2012, 5:131 Page 6 of 7 http://www.parasitesandvectors.com/content/5/1/131 small subunit ribosomal RNA sequences (18 S) of the twisted-wing parasites (Insecta: Strepsiptera). Insect Mol Biol 2005, 14:625–643.

37. Wilgenbusch JC, Swofford D: Inferring evolutionary trees with PAUP. Current Protocols in Bionformatics 2003, .
38. Petersen JJ, Willis OR: Experimental release of a mermithid nematode to control Anopheles mosquitoes in Louisiana. Mosquito News 1974, 34:316–319.
39. Becnel JJ, Johnson MA: Pathogenicity tests on nine mosquito species and several non-target organisms with Strelkovimermis spiculatus (Nematoda: Mermithidae). J Nematol 1998, 30:411–414.
40. Achinelly ME, Micieli MV, Marti GA, Garcia JJ: Susceptibility of neotropical mosquito larvae (Diptera: Culicidae) and non-target aquatic organisms to the entomoparasitic nematode Strelkovimermis spiculatus Poinar & Camino, 1986 (Nematoda: Mermithidae). Nematology 2004, 6:299–302.
41. Petersen JJ, Chapman HC: Parasitism of Anopheles mosquitoes by a Gastromermis sp. (Nematoda: Mermithidae) in southwestern Louisiana. Mosquito News 1970, 30:420–424.
42. Doucet MM, Laumond C, Bain O: Empidomermis riouxi nov sp (Nematoda: Mermithidae). Annales De Parasitologie Humaine Et Comparee 1979, 54:341–351.
43. Gaugler R, Wraight S, Molloy D: Bionomics of a mermithid parasitizing snow-pool Aedes spp. mosquitoes. Canadian J Zool 1984, 62:670–674.
44. St-Onge M, LaRue B, Charpentier G: A molecular revision of the taxonomic status of mermithid parasites of black flies from Quebec (Canada). J Invertebr Pathol 2008, 98:299–306.
45. Gillies MT, de Meillon B: The Anophelinae of Africa south of the Sahara (Ethiopian zoogeographical region). Johannesburg: South African Institute of Medical Research Publication; 1968.
46. Keiser J, Utzinger J, Castro MC, Smith TA, Tanner M, Singer BH: Urbanization in sub-Saharan Africa and implication for malaria control. AmJTrop Med Hyg 2004, 71:118–127.
47. Garrett-Jones C: Prognosis for interruption of malaria transmission through assessment of the mosquito's vectorial capacity. Nature 1964, 204:1173–1175.
48. Davies JB, McMahon JE, Beechgarwood P, Abdulai F: Does parasitism ofSimulium damnosum by Mermithidae reduce the

transmission ofonchocerciasis? Trans Roy Soc Trop Med Hyg 1984, 78:424–425. doi: 10.1186/1756-3305-5-131 Cite this article as: Kobylinski et al.: Mermithid nematodes found in adult Anopheles from southeastern Senegal. Parasites & Vectors 20125:131.

chapter 10

Isolation of Indigenous Strains of Paecilomyces lilacinus with Antagonistic Activity against Meloidogyne Incognita

Chen Guan Pau, Chan Teck Stephen Leong, Sing King Wong†, Lily Eng¶, Make Jiwan, Franklin Ragai Kundat, Zakry Fitri Bin AB. Aziz, Osumanu Haruna Ahmed, and Nik Muhammad Majid‡

Department of Crop Science, Faculty of Agriculture and Food Sciences, Universiti Putra Malaysia

†Department of Animal Science and Fishery, Faculty of Agriculture and Food Sciences, Universiti Putra Malaysia

‡Department of Forest Management, Faculty of Forestry, Universiti Putra Malaysia, Serdang, Malaysia

¶Agriculture Research Centre, Department of Agriculture Sarawak, Malaysia

ABSTRACT

Ten indigenous isolates of *Paecilomyces lilacinus* (PL), were isolated from two black pepper farms in Sarawak heavily infested with root-knot nematodes (RKN) as an initiative to control RKN problem. All isolates showed varying degree in colonizing female nematodes. In the female nematode bioassay on water agar, both indigenous strains of PL namely PLA, PLB, and a commercial strain, PLM (as positive control) demonstrated highly significant colonization (>90%, P≤0.01) on female. In egg parasitism test, spore suspension (10^5 spore/mL) of the strains PLA, PLB and PLM exhibited 78.8%, 66.0% and 73.4% parasitism on eggs, respectively. Meanwhile, hatching of nematode eggs incubated in spore suspension of PLA, PLB and PLM for seven days were significantly reduced; 88-89% of eggs were hatch-inhibited as compared to control (26%). This illustrated both local isolates, PLA and PLB are comparable with PLM as biological control agents for managing RKN infestation on black pepper vines.

INTRODUCTION

Black pepper (*Piper nigrum* L.) is an important cash crops of Sarawak, with 6,125 tonnes of black pepper exported from January to June 2010, valued at RM77.889 million (Anonymous, 2010). However, most of black pepper farms in Sarawak are infested by *Meloidogyne spp*. (Kueh, 1978; Leong, 1986; Ravindran, 2000; Eng, 2001), which has become a major constraint to the black pepper production. Survey conducted by Eng (2001) in 43 black pepper farms revealed that root-knot nematodes (RKN) were present in all the farms, implying that black pepper is one of the important hosts of RKN and the problem of RKN infestation in these farms. Currently, no resistant cultivar to RKN is available (Eng, 2001). For the past decades, more attention have been directed to the application of microbial control agents since chemical nematicides are being reappraised in respect of health and environmental concern and limited availability in developing nations (Mukhtar & Pervaz, 2003; Dong & Zhang, 2006).

Paecilomyces lilacinus (PL), a saprophytic soil fungus has drawn many research attentions due to its promising effect in parasitizing and controlling population of phytonematodes (Jatala, 1986; Dube & Smart, 1987; Hewlett *et al.*, 1988; Freitas *et al.*, 1995; Nagesh *et al.*, 1997; Khan *et al.*, 2006a; Kiewnick & Sikora, 2006; Brand *et al.*, 2010). It has a high frequency of occurrence in the tropics and subtropic (Morgan *et al.*, 1984; Chen *et al.*, 1996) and can be found in most of agricultural soils (Brand *et al.*, 2010). It has been recognized as a common egg pathogenic fungus of root-knot and cyst nematodes (Rumbos & Kiewnick, 2006). In the study of Eng (2001), 82.9% of the 41 surveyed farms in Sarawak were reported to contain PL despite intensive application of fungicide in the black pepper holdings. Besides, PL has high adaptability

in its life strategy, making it competitive in a broad spectrum of range adaptability. It can tolerate wide range of soil pH and able to grow well at 15 -30°C. Cabanillas *et al.*

(1989) observed maximum growth of PL at temperature from 24-30°C and reported its' ability to grow and compete for a wide range of common substrate in soil. The present study aimed to isolate local virulent PL strains and to evaluate their *in vitro* antagonistic activity against different stage of RKN. This may provide opportunity to further evaluate their efficacy in pot and field trial.

MATERIALS AND METHODS

Establishment of Pure Nematode Culture

A pure culture of *M. incognita* was established by a single egg mass and maintained on tomato plants (*Lycopersicon esculentum*) raised in pot filled with sterilized pot mixture (2: 1: 1 sand, loam, cocoa peat respectively).

Isolation, Screening and Identification of PL

Seven strains of PL, designed as PLA, PLEJ1, PLEJ2, PLEK1, PLEK2, PLEK3 and PLEK4 were isolated from egg masses and female nematodes while three strains, designed as PL1A, PLB and PLSA were obtained from soil. PLM, a commercial strain (SUBOTANITM) served as positive control in all tests. Roots and rhizosphere soils around black pepper roots were collected from several areas in two black pepper farms (UPM Campus Bintulu, N03°13′00″ E 113°05′56″ & Tondong, N01°26′59″E110°08′2″), where RKN disease is

prevalent. Root pieces were washed in gentle running tap water for 5 min. Females and egg masses were extracted from roots using method described by Sun *et al.* (2006) before transferred to PDA+ [Potato Dextrose Agar (Merck, Darmstadt, Germany) amended with 0.01% (w/w) Chloramphenicol (Sigma, China) and 3%(w/w) Sodium Chloride] plate. For isolation from soil: Serial dilution and pour plate technique was used (Johnson & Curl, 1972). Dilution at 10-1, 10-2 and 10-3, were subsequently spread onto PDA+ plates and incubated at room temperature (28 ± 1°C) for 7 days. Identification of PL was based on cultural and morphological characters (Samson, 1975). Axenic cultures were obtained by single spore isolations using fine needle (Johnston & Booth, 1983) and maintained on PDA slants. PLA, PLB and PLM that exhibited highest colonization on female in the later test were further species confirmed at molecular level with primer pairs: ITS1-ITS4 (Inglis & Tigano, 2006) and EF4-EF3 (Glass & Donaldson, 1995) before being assessed for their *in vitro* antagonism on egg mass, eggs, egg hatch and Juvenile (J2).

PL Mass Production

Sterilized substrate (composition: 150 g rice husk, 60 mL distilled water, 10% (w/w) molasses, pH 6.5) packed in polyethylene bag (45 mm x 35 mm) was inoculated with 1 x 108 PL conidia. The top of the bag was stuffed up into a cotton-filled piece of PVC pipe (3-cm-diameter & 5-cm-length) that was vertical to the bag. The bag was incubated in the dark at room temperature with periodical shaking (5 min every morning) for aeration and even growth. After 14 days, 5 g of fermented substrate was collected in a sterile test tube, added with 10 mL of 0.05% Tween 80, vortexed for 2 min to dislodge and suspend the spore before being filtered through a doubled layer sterile cheese cloth. Spore concentration was determined using a haemocytometer and was later standardized to 105 spores/mL.

Female Nematodes Bioassay

PLA, PLEJ1, PLEJ2, PLEK1, PLEK2, PLEK3, PLEK4, PL1A, PLB, PLSA and PLM were preliminary screened for their efficacies in colonizing female nematodes. Five equal size females (from pure culture) were extracted from tomato roots, surface sterilized as described before placed at the edge of 5-d-old fungal colony grown on water agar 2% (w/w). Plates were run in triplicate, arranged in a completely randomized design (CRD) and incubated at room temperature. After 4 days, females were observed under stereomicroscope (45x) to detect emerging mycelial from the body surface as sign of colonization.

Colonization R*ate* on Female and Egg Mass

Colonization rate on female by isolate PLA, PLB and PLM was conducted with ten female per plate and four replicate plates per fungus treatment. Sign of colonization was observed daily until the 6th day. Similarly, the procedure was repeated for egg mass.

PL Parasitism on Eggs

Egg suspension was prepared as described by Nitao *et al.* (1999). A 50 μL of a freshlyn prepared egg suspension (approximately 300 fresh eggs) was pipetted into a sterile McCaney bottles and added with 1 mL of fungus spore suspension (105 spore/mL). Sterile water was added for control treatment. The entire procedure was conducted in aseptic condition. Bottles were sealed with parafilm, arranged in CRD and placed in the dark at room temperature with four replicates per fungus treatment. Four days later, drops of egg suspension from each replicate (approximately 100 eggs) were pipetted onto glass slides, stained with lactophenol cotton blue and examined under a compound microscope (400x) for sign of parasitism.

Effect of Spore Suspension on Egg Hatch and Juvenile (J2)

Procedure described in egg parasitism test was repeated with fresh material but incubation period was extended to seven days. Egg hatch rate and J2 mortality were determined by counting all eggs, J2 and dead J2 in a counting disc under a stereo-microscope and calculated according to the following formula:

$$\text{Egg hatch rate} = 100 \times J2/(\text{eggs} + J2)$$

$$\text{Juveniles mortality} = 100 \times \text{dead } J2/\text{total } J2.$$

A piece of eyelash attached to a tooth pick was used to probe the tail of J2. J2 were considered dead if they became rigid and did not react when probed by the eyelash.

Statistical Analysis

Means of data in percentage were subjected to arc-sine transformation, analyzed according to standard procedure for analysis of variance (ANOVA). Differences between means were compared using SAS version 9.0 for significance according to Duncan multiple range test ($P<0.01$ & $P<0.05$). Untransformed arithmetic means are reported.

RESULTS AND DISCUSSION

Isolation of PL

Growth of colony on PDA plate formed a basal felt with floccose aerial mycelium giving rise to conidiophores. Diameter of colonies ranged 5-7 cm within 14 days incubation at 28°C ± 2. Conidial heads appeared white but gradually became light brownish when sporulated (Fig. 1). This colour change is consistent with species of PL reported by Samson (1975). The incorporation of 3% NaCl to PDA gave a suppressive effect on other fungal species, while maintaining the growth and sporulation of PL. Chloramphenicol, a heat stable antibiotic was added to kill a wide range of Gramnegative and Gram-positive bacteria.

Microscopic observation (1000x) revealed conidiophores arise from aerial hyphae, have rough wall, bore verticillate short branches with whorls of 2-4 phialides. The hyphae was hyaline and bear flask shaped hialides (P) with huge numbers of conidia (C) attached loosely in long divergent chains (Fig. 2). The phialide possessed a swollen basal and a distinct neck. Conidia were in the shape of ellipsoidal to fusiform (Samson, 1975).

Table 1: Effect of PL spore suspension (10^5 spore/mL) on egg parasitism, egg hatch inhibition and J2 mortality

Isolate	PLA	PLB	PLM	Control
Parasitized egg (%)	78.5±2.4^a	66.0±5.2^b	73.4±0.6^{ab}	0.0±0^c
Hatch Inhibited egg (%)	88.3±1.4^a.	88.2±1.7^a	89.4±1.0^a	25.9±3.12^b
Dead J2 (%)	6.0%±1.8^a	5.5%±0.8^a	5.7%±1.0^a	2.3%±0.8^b

Each value (%) represents the mean of four replicates ± standard error. Values within a raw followed by the same letter are not significantly different according to Duncan Multiple Range Test at P=0.05

Figure 1: Growth morphology of PL colonies on PDA agar plate.

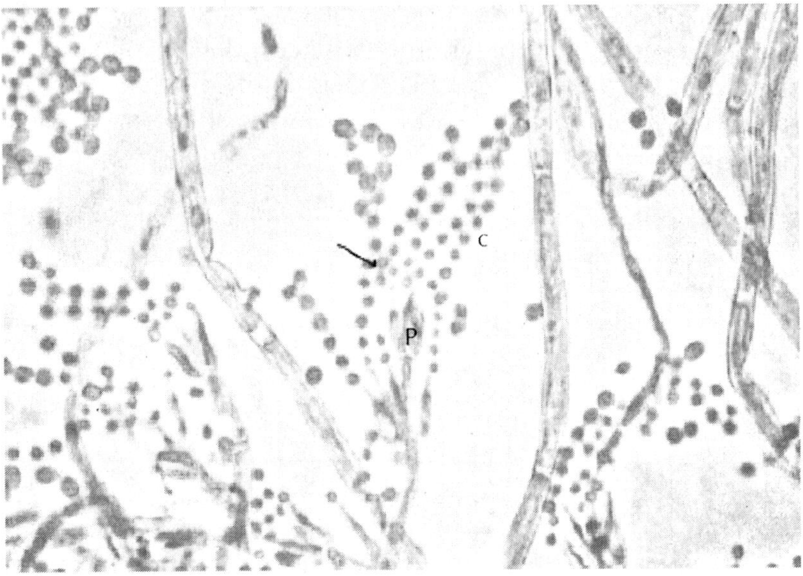

Figure 2: Hyphae bear phialides (P) with huge numbers of conidia (C) attached loosely in long divergent chains.

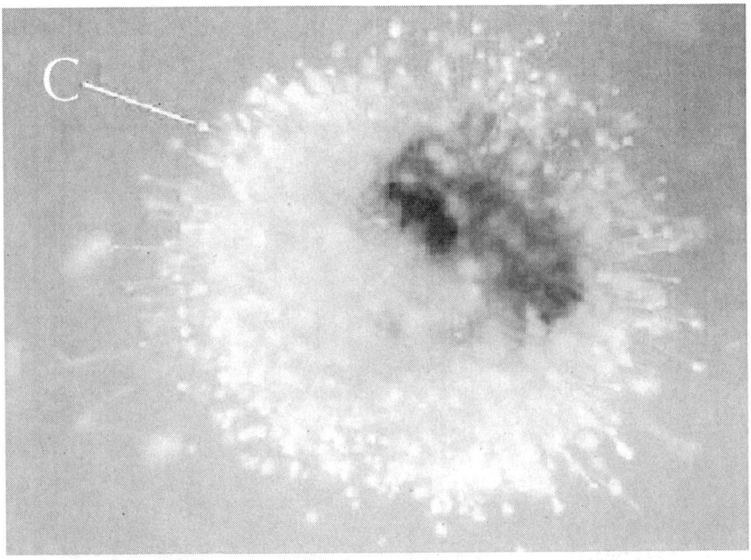

Figure 3: Colonization on female nematode by commersial strain PLM. Hundreds of conidiosphores (C) radiating from the body surface.

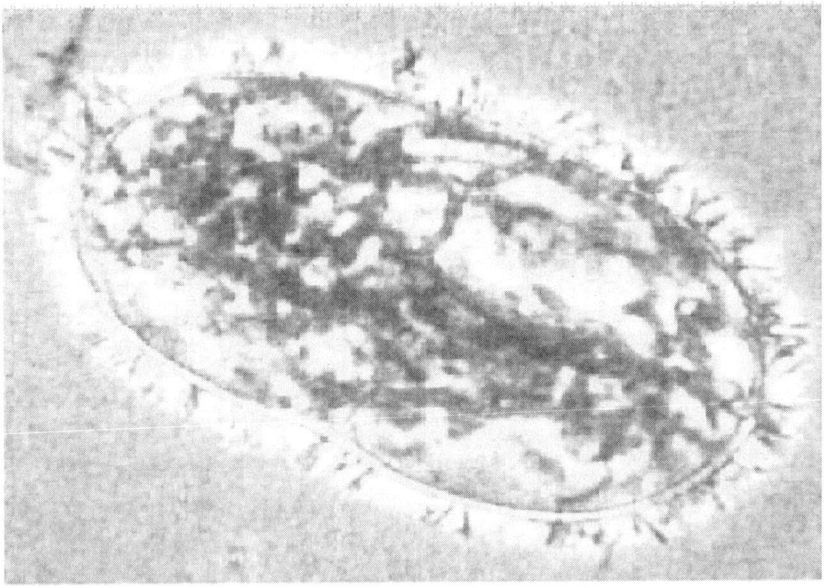

Figure 4: Hyphae emerged from a deformed shape egg with disintegrated embryo after 4 days of incubation (400 x). It also penetrated into the egg and consumed the egg content.

Female Nematodes Bioassay

All PL isolates (Fig. 5) showed varying degrees in colonizing female with PLA, PLB, PLEJ1 and PLM demonstrated the highest significant ($P<0.01$) colonization (>90%), which is consistent with the result of Eapen *et al.* (2005) on the significant parasitism of their PL strain on female nematode. Observation under a stereomicroscope (40x) demonstrated radiating mycelia from the female's body (Fig. 3) unrestricted to vulva, anus or broken opening, which is in concert with the reports of Morgan-Jone *et al.* (1984), Holland *et al.* (1999) and Khan *et al.* (2006b). They suggested that appressoria are not involved in the penetrating process owing to the lack of a tough crosslinked chitin layer on the female body. However, Jatala (1986) reported a contradictory observation that Peruvian isolate of PL infected female of *M. javanica* only by body opening. Khan *et al.* (2006b) explained that the difference could be due to different profile enzymes of each strain that help in the direct cuticle penetration.

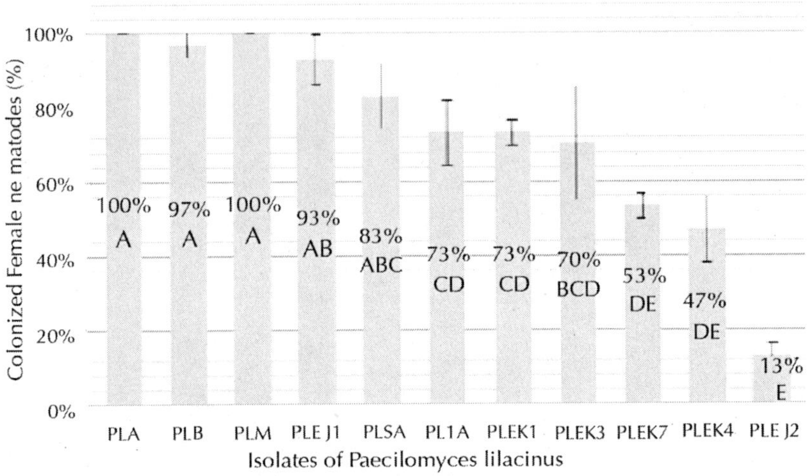

Figure 5: Percentage of female nematodes colonized by PL. Value in each bar represents the mean (%) of three replicates. Bars marked by the same letter are not significantly different according to Duncan Multiple Range Test at P≤0.01

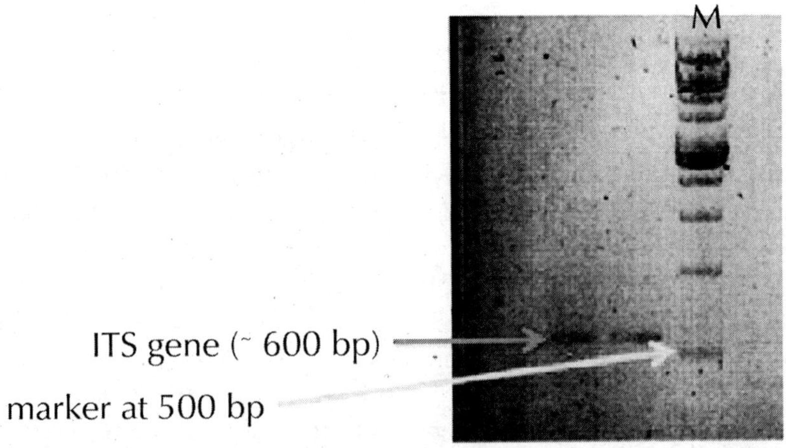

Figure 5a: Detection of PCR amplification of ITS gene. M represents VC 1kb DNA ladder (Vivantis). For primers ITS1 and ITS4, the size of amplified PCR product was approximately 450-800 bp

Molecular Identification of PL

For primers EF3 and EF4, the size of amplified PCR product was approximately 1.4-1.5 kb whereas for primers ITS1 and ITS4, the size was approximately 450-800 bp (Fig. 5a). Sequence analysis of PLA clone showed 95-99% similarity to PL sequence in the GenBank, EMBL, DDBJ, PDB database while sequence of PLB clone showed 94% similarity. This molecularly confirmed that both PLA and PLB isolates are of the species PL.

Colonization Rate on Female and Egg Mass

Isolate PLA, PLB and PLM did not demonstrate high significant colonization rate on female (Fig. 6). On the 1st day, PLA and PLM demonstrated 50% infection rate followed by PLB with rather low rate (10%). However, all isolates had achieved complete colonization on the 6th day.

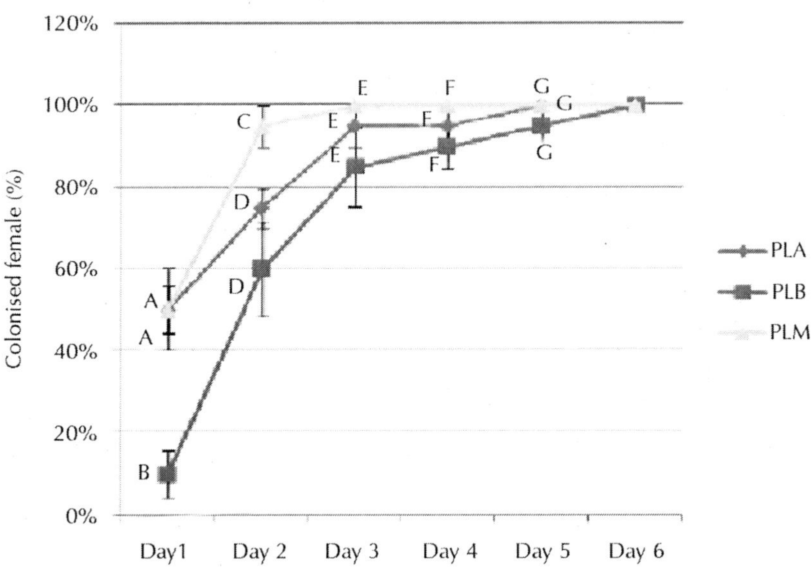

Figure 6: Percentage of female nematodes colonized by

PL under *in vitro* condition. Dots on each graph line represent mean (% of colonized female) of 4 replicates, marked by the same letter for each respective day are not significantly different according to Duncan Multiple Range Test at P≤0.05. Error bar represent standard error

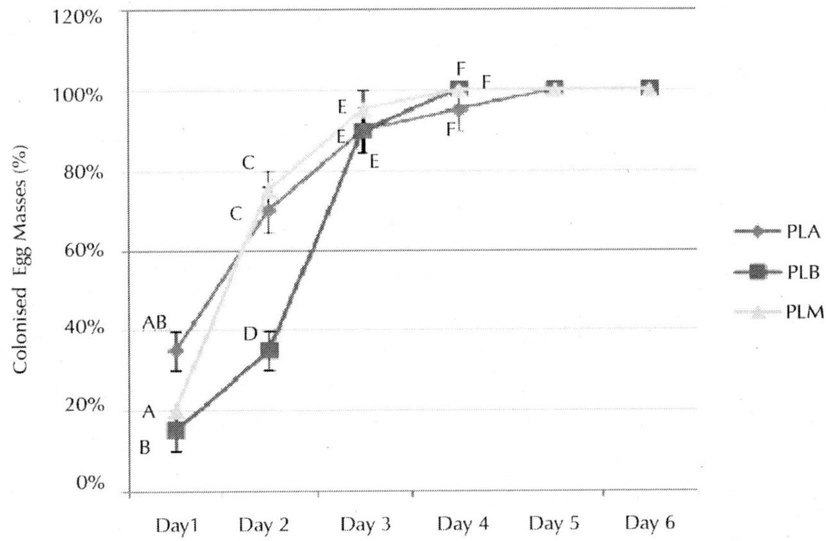

Figure 7: Percentage of egg masses colonized by PL

under *in vitro* condition. Dots on each graph line represent mean (% of colonized egg mass) of 4 replicates, marked by the same letter for each respective day are not significantly different according to Duncan Multiple Range Test at P≤0.05. Error bar represent standard error

Colonization on egg masses by PLA, PLB and PLM was confirmed by observing the emerging mycelial from the surface of egg masses under a stereomicroscope 40X colonization rate of 15-35% on egg masses, but drastically increased to 90-95% on the 3rd day and achieved complete colonization on the 5th day. Similarly, all the three isolates showed insignificant colonizing rate on egg mass. The presence of antimicrobial compound (Orion *et al.*, 2001) in gelatinous matrix (GM) of an egg mass protects eggs from microbial infection (Kiewnick *et al.*, 2006). Sharon *et al.* (2007) claimed that an effective parasite should be able to utilize GM as source of nutrient and reproduce in it.

Since mycelial was detected on the GM, it is suggested that PLA and PLB possess resistance attribute to antimicrobial compound in the GM, which is in line with those of Zaki and Bhatti (1990) and Eapen *et al.* (2005).

PL Parasitism on Eggs

PLA, PLM and PLB demonstrated high and significant parasitism on eggs as compared to control but not significant among themselves, 78.5%, 73.4% and 66.0%, respectively (Table I). This result is in conformity with Sun *et al.* (2006) who reported a high *in vitro* parasitism rate of PL strain YESX- 2-14 on *M. hapla* egg; also Al Kader (2008) who revealed 77% eggs infection after 4 days of incubation. In addition, thirty PL isolates were reported to parasitize 100% of *M. hapla* eggs (Sun *et al.*, 2006).

In this study, early age eggs were more susceptible to PLA and PLB infection than eggs with ready to hatch J2. In most cases, infected eggs contained no J2 with egg content and embryo seemed disintegrated (Fig. 4). This suggested that most immature eggs were parasitized but some eggs containing developing J2 were also infected, and J2 seemed motionless. This is similar to the studies of Irving and Kerry (1986), Jatala (1986), Lopez-Llorca and Duncan (1991) and Eapen *et al.* (2005). However, Morgan-Jone (1984) and Holland *et al.* (1999) claimed that their PL strain infected eggs at all developing stages. This attribute offers advantage in biological control since eggs are the major target of parasitic fungus and can remain dormant in soil for long periods of time (Holland *et al.*, 1999).

Under microscopic observation, extensive network of hyphae were detected ramifying several eggs but not growing on other adjacent eggs in the group. At the end of hyphae, a simple swollen hyphae structure, recognized as appressorium appeared pressing onto the eggshell. Some incubated eggs appeared abnormal, deformed and shrunken owing to the pressure exerted by the network of hyphae. This indicates the use of mechanical means to penetrate host, which has been described by Holland *et al.* (1999). When a hyphae encounters an egg surface, it responds thigmotrophically by forming appressoria (Lopez-Llorca *et al.*, 2002), followed by adhesion of an adhesive (an extracellular material on appressoria) to the egg for better binding of fungus to the host (Lopez-Llorca *et al.*, 2008). From these appressoria, the fungus use enzymatic and mechanical means to penetrate the host (Morgan-Jones *et al.*, 1984; Huang *et al.*, 2004; Gortari & Hours, 2008; Lopez-Llorca *et al.*, 2008). According to Morton *et al.* (2004), the secretion of chitinase and proteases by PL facilitated egg penetration by breaking down layers (the barrier) in egg shells so that a narrow infection peg can push through. The combined effect of chitinase and protease produced by PL in degrading eggshell layers of *M. javanica* was proven by Khan *et al.*

(2004). The lipid layer disappeared, while the chitin layer was much reduced after enzyme treatment.

Effect of Spore Suspension on Egg Hatch Inhibition

Egg hatch was not significantly different among the three isolates but significant ($P< 0.05$) with the control 25.9% (Table I). PLM recorded the highest inhibition at (89.4±1.0%), followed by PLA at (88.3±1.4%) and PLB at (88.2±1.7%). Eggs treated with spore suspension appeared shrunken, deformed and with multiple vacuoles. Most eggs in control treatment appeared empty, with many J2 appeared outside the eggs, indicating hatching had occurred. This result is in agreement with Sun *et al.* (2006) who reported average 58% egg hatch inhibition for their 186 PL isolates. Costa *et al.* (2001) reported that culture filtrate of PL grown in Czapek broth greatly reduced egg hatching of MI. According to Bonants *et al.* (1995), hatching of eggs containing mature J2 appeared to be stimulated when incubated in culture filtrate of PL but development of immature eggs appeared to be disrupted.

Effect of Spore Suspension on J2 Mortality

To infect J2, PL needs to overcome the cuticle of nematode. The cuticle is a non-cellular layer production of the hypodermis, which consists of keratin, collagen and fibers (Huang *et al.*, 2004). Once the cuticle is penetrated by fungi hyphae, the nematode get paralyzed, invaded and digested (Tunlid & Jansson, 1991). In this study, spore suspension of PLA, PLB and PLM, demonstrated low mortality effect on J2, 6.0%, 5.5% and 5.7%, respectively with no significant difference ($P< 0.05$).

This confirmed the findings of Sun *et al.* (2006) who reported a low average J2 mortality percentage of 16% for their 186 PL isolates. Also, it agrees with the reports of Jatala (1986), Bonants *et al.* (1995), Singh and Mathur (2010) that PL primarily colonized eggs but not juveniles. In contrast, PL strain YES-X-2-14 was reported in exhibiting high *in vitro* nematicidal effect on J2 of *M. hapla* (Sun *et al.,* 2006). Similarly, Al Kader (2008) reported a high nematicidal effect of their PL culture filtrate on J2 of MI, with 99% of J2 immobilized after 2 days of treatment. It is suggested that different strain of PL exhibit different nematicidal effect on J2. Throughout the years, metabolites in culture filtrate of PL have been detected and screened such as paecilotoxin (Mikami *et al.*, 1989), acetic acid (Djian *et al.*, 1991) and leucinostatins (Park *et al.*, 2004). These metabolites may potentially cause nematicidal effect toward J2 of nematodes. Culture filtrate of local isolate PLA and PLB grown in PDB recorded acidic pH of 3.30-3.46 (data not shown). It is suspected that both local isolates produced acetic acid, thus proposing a further study to be performed to screen and confirm the active compounds present in the culture filtrate.

ACKNOWLEDGEMENTS

The authors would like to thank Sharon Wong and Lau Sien Loong for their technical assistance in PCR. Great appreciation is also extended to Malaysian Pepper Board for funding this project, Fund no: 6300301-11601.

REFERENCES

1. Al Kader, M.A.A., 2008. *In vitro* studies on nematode interaction with their antagonistic fungi in the rhizosphere of various plants. *Ph.D. Thesis*, p: 58. Albert-Ludwigs-Universität, Germany
2. Anonymous, 2010. Market Review January-June 2010. *In*: Rapaee, T.H.S. (ed.), *Malaysia Pepper Industry Bullentin Jan-Jun 2010*, pp: 10–16. Malaysian Pepper Board, Kuching Sarawak
3. Brand, D., C.R. Soccol, A. Sabu and S. Roussos, 2010. Production of fungal biological control agents through solid state

fermentation: A case study on *Paecilomyces lilacinus* against root-knot nematodes. *Micol. Aplicada Int.,* 22: 31–48

4. Bonants, P.J.M., P.F.L. Fitters, H. Thijs, E.D. Belder, C. Waalwijik and J.W.D.M. Henfling, 1995. A basic serine protease from *Paecilomyces lilacinus* with biological activity against *Meloidogyne hapla* eggs. *Microbiology,* 141: 775–784
5. Cabanillas, E. and K.R. Barker, 1989. Impact of *Paecilomyces lilacinus* inoculums level and application time on control of *Meloidogyne incognita* on tomato. *J. Nematol.*, 21: 115–120
6. Chen, Z.X., D.W. Dickson, R. McSorley, D.J. Mitchell and T.E. Hewlett, 1996. Suppression of *Meloidogyne arenaria* race 1 by soil application of endospores of *Pasteuria penetrans. J. Nematol.*, 28: 159–168
7. Costa, M.J.N., V.P. Campos, L.H. Pfenning and D.F. Oliveira, 2001. Toxicity of filtered fungals to *Meloidogyne incognita. Fitopatol. Brasilia*, 26: 749–755
8. Djian, C., L. Pijarowski, M. Ponchet, N. Arpin and J. Favre-Bonvinm, 1991. Acetic acid: a selective nematicidal metabolite from culture filtrates of *Paecilomyces lilacinus* (Thom) Samson and *Trichoderma longibrachiatum* Rifai. *Nematologica,* 37: 101–112
9. Dong, L.Q. and K.Q. Zhang, 2006. Microbial control of plant-parasitic nematodes: a five-party interaction. *Plant Soil*, 288: 31–45
10. Dube, B. and G.C. Jr2. Smart, 1987. Biological control of *Meloidogyne incognita* by *Paecilomyces lilacinus* and *Pasteuria penetrans. J. Nematol.,* 19: 222–227
11. Eapen, S.J., B. Beena and K.V. Ramana, 2005. Tropical soil microflora of spice-based cropping systems as potential antagonists of rootknot nematodes. *J. Invertebr Pathol.,* 88: 218–225
12. Eng, L., 2001. Biological control of Root-knot nematodes (*Meloidogyne* species) on Black Pepper (*Piper nigrum* L.) in Sarawak. *Ph.D. Thesis*, University of Reading, United Kingdom
13. Freitas, L.G., S. Ferraz and J.J. Muchovej, 1995. Effectiveness of different isolates of *Paecilomyces lilacinus* and an isolate of

Cylindrocarpon destructans on the control of *Meloidogyne javanica*. *Nematropica*, 25: 109–115

14. Glass, N.L. and G.C. Donaldson, 1995. Development of primers set designed for use with the PCR to amplify conserved genes from filamentous Ascomycetes. *Appl. Environ. Microbiol.,* 61: 1323–1330
15. Gortari, M.C. and R.A. Hours, 2008. A review: Fungal chitinases and their biological role in the antagonism onto nematode eggs. *Mycol. Prog.*, 7: 221–238
16. Hewlett, T.E., D.W. Dickson, D.J. Mictchell and M.E. Kannwischer-Mitchell, 1988. Evaluation of *Paecilomyces lilacinus* as a biocontrol agent of *Meloidogyne javanica* on tobacco. *J. Nematol.,* 20: 578–584
17. Holland, R., K.L. Williams and A. Khan, 1999. Infection of *Meloidogyne javanica* by *Paecilomyces lilacinus*. *Nematology*, 2: 131–139
18. Huang, X., N. Zhao and K. Zhang, 2004. Extracellular enzymes serving as virulence factors in nematophagous fungi involved in infection of the host. *Res. Microbiol.,* 155: 811–816
19. Inglis, P.W. and M.S. Tigano, 2006. Identification and taxonomy of some entomopathogenic *Paecilomyces* spp. (Ascomycota) isolates using rDNA-ITS Sequences. *Genet. Mol. Biol.*, 29: 132–136
20. Irving, F. and B.R. Kerry, 1986. Variation between strains of the nematophagous fungus, *Verticillium chlamydosporium* Goddard. II. Factors affecting parasitism of cyst nematode eggs. *Nematologica*, 32: 474–485
21. Jatala, P., 1986. Biological control of plant parasitic nematodes. *Ann. Rev. Phytopathol.*, 24: 453–489
22. Johnson, L.F. and E.A. Curl, 1972. *Methods of Research on the Ecology of Soil Borne Plant Pathogens*, pp: 6–8. Burges Publishing Company, Minnesota, USA
23. Johnston, A. and C. Booth, 1983. *CMI Plant Pathologist Pocket Book*, p: 439. Common wealth Agriculture Bureaux, UK
24. Khan, A., K.L. Williams and H.K.M. Nevalainen, 2006a. Control of plant-parasitic nematodes by *Paecilomyces lilacinus* and *Monacrosporium lysipagum* in pot trials. *Biol. Cont.*, 51: 643–658

25. Khan, A., K.L. Williams and H.K.M. Nevalainen, 2006b. Infection of plant-parastic nematodes by *Paecilomyces lilacinus* and *Monacrosporium lysipagum*. *Biol. Cont.*, 51: 659–678

26. Khan, A., K.L. Williams and H.K.M. Nevalainen, 2004. Effects of *Paecilomyces lilacinus* protease and chitinase on the eggshell structures and hatching of *Meloidogyne javanica* juveniles. *Biol. Cont.*, 31: 346–352

27. Kiewnick, S., A.J. Lauritzen and J. Eilenberg, 2006. Investigations on the colonization of root-knot nematode galls and egg masses by the antagonistic fungus *Paecilomyces lilacinus* strain 251. *Phytopathology*, 96: 60

28. Kiewnick, S. and R. Sikora, 2006. Evaluation of *Paecilomyces lilacinus* strain 251 for the biological control of the northern root-knot nematode *Meloidogyne hapla* Chitwood. *Nematology*, 8: 69–78

29. Kueh, T.K. and C.H. Teo, 1978. Chemical control of root-knot nematodes in *Piper nigrum*. *Planter*, 54: 37–245

30. Leong, C.T.S., 1986. *Pepper Nematodes. In: Annual Report for the Year 1984*. Department of Agriculture, Sarawak, Borneo

31. Lopez-Llorca, L.V. and G.H. Duncan, 1991. Effects of fungal parasites on cereal cyst nematodes (*Heterodera avenae* Woll.) from naturally infested soil–a scanning electron microscopy study. *Canadian J. Microbiol.*, 37: 218–225

32. Lopez-Llorca, L.V., J.G. MaciÁ-Civente and H.B. Jansson, 2008. Mode of action and interactions of nematophagous fungi. *In:* Ciancio, A. and K.G. Mukerji (eds.), *Integrated Management and Biocontrol of Vegetable and Grain Crops Nematodes*, pp: 51–76. Springers, The Netherlands

33. Lopez-Llorca, L.V., C. Olivares-Bernabeu, J. Salinas, H.B. Jansson and P.E. Kolattukudy, 2002. Prepenetration events in fungal parasitism of nematode eggs. *Mycol. Res.*, 106: 499–506

34. Mikami, Y., K. Yazawa, K. Fukushima, T. Arai, S.I. Udagawa and R.A. Samson, 1989. Paecilotoxin production in clinical or terrestrial isolates of *Paecilomyces lilacinus* strains. *Mycopathologia,* 108: 195–199

35. Morgan-Jones, G., J.F. White and R. Rodriguez-Kabana, 1984. Phytonematode pathology: Ultrastructural studies II. Parasitism of

Meloidogyne arenaria eggs and larvae by *Paecilomyces lilacinus*. *Nematropica*, 14: 57–71

36. Morton, C.O., P.R. Hirsch and B.R. Kerry, 2004. Infection of plantparasitic nematodes by nematophagous fungi –a review of the application of molecular biology to understand infection processes and to improve biological control. *Nematology*, 6: 161–170

37. Mukhtar, T. and I. Pervaz, 2003. *In vitro* evaluation of ovicidal and larvicidal effects of culture filtrate of *Verticillium chlamydosporium* against *Meloidogyne javanica*. *Int. J. Agric. Biol.*, 5: 576–579

38. Nagesh, M., P. Parvatha Reddy and M.S. Rao, 1997. Integrated management of *Meloidogyne incognita* on tuberose using *Paecilomyces lilacinus* in combination with plant extracts. *Nematol. Medit.*, 25: 3–7

39. Nitao, K.N., S.L.F. Meyer and D.J. Chitwood, 1999. *In vitro* assays of *Meloidogyne incognita* and *Heterodera glycines* for detection of nematode-antagonistic fungal compounds. *J. Nematol.*, 1: 172–183

40. Orion, D., G. Kritzman, S.L.F. Meyer, E.F. Erbe and D.J. Chitwood, 2001. A role of the gelatinous matrix in the resistance of root-knot nematode (*Meloidogyne* spp.) eggs to microorganisms. *J. Nematol.*, 33: 203–207

41. Park, J.O., J.R. Hargreaves, E.J. McConville, G.R. Stirling, E.L. Ghisalberti and K. Sivasithamparam, 2004. Production of leucinostatins and nematicidal activity of Australian isolates of *Paecilomyces lilacinus* (Thom) Samson. *Lett. Appl. Microbiol.*, 38: 271–276

42. Ravindran, P., 2000. *Black Pepper Piper nigrum*. Harwood academic publishers, The Netherlands

43. Rumbos, C. and S. Kiewnick, 2006. Effect of plant species on persistence of *Paecilomyces lilacinus* strain 251 in soil and on root colonization by the fungus. *Plant Soil*, 283: 25–31

44. Samson, R.A., 1975. *Paecilomyces and some Allied Hyphomycetes: Studies in Mycology*, Vol. 6, pp: 1–119 Centraalbureau voor Schimmelcultures, Baarn, The Netherlands

45. Sharon, E., I. Chet, A. Viterbo, M. Bar-Eyal, H. Nagan, G.J. Samuels and Y. Spiegel, 2007. Parasitism of *Trichoderma* on *Meloidogyne*

javanica and role of the gelatinous matrix. *European J. Plant Pathol.,* 118: 247–258

46. Singh, S. and N. Mathur, 2010. *In vitro* studies of antagonistic fungi against the root-knot nematode, *Meloidogyne incognita*. *Biocontrol Sci. Technol.*, 20: 275–282
47. Sun, M.H., L. Gao, Y.X. Shi, B.J. Li and X.Z. Liu, 2006. Fungi and actinomycetes associated with *Meloidogyne spp.* eggs and in China and their biocontrol potential. *J. Invert. Pathol.*, 93: 22–28
48. Tunlid, A. and S. Jansson, 1991. Proteases and their involvement in the infection and immobilization of nematodes by nematophagous fungus *Arthrobotrys oligospora*. *Appl. Environ. Microbiol.*, 57: 2868–2872
49. Zaki, P.A. and D.S. Bhatti, 1990. *In vivo* parasitism of *Meloidogyne javanica* by an oviparasitic fungus, *Paecilomyces lilacinus*. *Nematol. Medit.*, 18: 141–143

Chapter 11

Evaluation of Stropharia sp. 1.2052 Nematicidal Effects against Meloidogyne Incognita on Tomato

Chuixu Kong, Chongyan Zhai, Jing Li, Jun Zhang, Keqin Zhang and Yajun Liu

Laboratory for Conservation and Utilization of Bio-resources and Key laboratory of Microbial Diversity, Southwest China, Ministry of Education, Yunnan University, Kunming 650091, P. R. China.

ABSTRACT

To use unique nematicidal mechanism of *Stropharia*, the strain 1.2052 was evaluated for its potential to control the root-knot nematode *Meloidogyne incognita* on tomato. *In vitro*, the inhibition rate in 24 h of *Stropharia* sp. 1.2052 isolate was 100% for second-stage juveniles of *M. incognita* and *Panagrellus redivivus*, 41.81% for *Caenorhabiditis elegans* and 99.25% for *Bursaphelenchus xylophilus*, respectively. In pot experiments, isolate 1.2052 reduced the root-knots to 68.16

to 84.19% in one month and 45.28 to 88.24% in two months after treatment, respectively. The reduction of nematodes in soil ranged from 26.39 to 61.18% compared to the negative control. There were significant efficacies for control of root knot nematodes on tomato.

INTRODUCTION

Root-knot nematodes (*Meloidogyne* spp.) cause serious damages to agriculture and forest. It has been reported that plant parasitic nematodes cause annual losses of several billion of crops worldwide (Koenning et al., 1999; Perry et al., 2009). Various fungi have been tested and used to control root-knot nematodes during the last decades. The most studied fungi used and applied to control nematodes are *Hirsutella rhossiliensis* and *Pochonia chlamydosporia*. *H. rhossiliensis* is one of few endoparasitic fungi of mobile stage of nematode. Both species produce infective, adhesive conidia that may be attached to the cuticle of a passing nematode, and has shown great potential in soybean cyst nematode control (Li et al., 2008). *P. chlamydosporia* has a worldwide distribution, and has been found in nematode suppressive soils to parasitize nematode eggs (Manzanilla-Lo´pez et al., 2013). *Fusarium oxysporum* f. sp. *glycines* and *Sclerotium rolfsii* were studied for their effects on pathogenicity of *M. incognita* race 2 in soybean (Akinsanmi et al., 2003). Kiewnick et al. (2006) reported that a *Paecilomyces lilacinus* strain was selected to control *M. incognita*. *Arthrobotrys dactyloides* was tested against *M. incognita* (J2) on Tomato (Kumar et al., 2006). *Muscodor albus* was evaluated as a potential biocontrol agent against four plant-parasitic nematodes of econo-mically important vegetable crops (Riga et al., 2008). Biological control of root-knot nematode (*M. javanica)* by *Trichoderma harzianum* BI was also investigated in greenhouse and laboratory experiments (Sahebani and Hadavi, 2008).

Species of *Stropharia* (Strophariaceae, Basidiomycota) have purple-brown to tobacco brown spores and appear in woods, grasslands, compost piles, and animal dung. *Stropharia* cultures can produce unique stellate cells, called acanthocytes. Farr (1980) has made a detailed study of the morphology and development of acanthocytes, and considered the ability to form acanthocytes as a characteristic of the whole genus (Norvell et al., 2000). Luo et al. (2006) reported that *S. rugosoannulata* showed ability to immobilize the nematodes

Panagrellus redivivus and Bursaphelenchus xylophilus within hours on agar plates.

The nematode-attacking activity of this fungus is carried out by spiny acanthocytes and their mechanical force is an important factor in the process. Furthermore, the growth and nematode-attacking activity of the fungus in soil were also determined, suggesting that acanthocytes are functional in soil. In the present study, mechanism of *Stropharia* sp. 1.2052 activities against a population of the root-knot nematode *M. incognita* was evaluated on tomato, in greenhouse experiments.

MATERIALS AND METHODS

Fungal Culture

Stropharia sp. 1.2052 was collected from forest in Yunnan, P. R. China. The pure culture was isolated from fruiting body, identified by Luo et al. (2006), and deposited in the culture collection of Key Laboratory for Conservation and Utilization of Bio-resource, Yunnan University, P. R. China.

Solid culture of this strain was prepared on wheat grains. Wheat was soaked 24 h with tap water and 1 h with boiling water, later mixed with CaCO3, glucose and sawdust (wheat, 60%; sawdust, 36%; CaCO3, 2%; glucose, 2% v%). Such culture medium was transferred in 250 ml flask and sterilized at 121°C for 2 h. Inoculated cultures were incubated at 25°C for 30 days when more than 80% wheat was consumed.

Nematodes Culture

M. Incognita

The nematodes population was collected from tobacco in Yunnan, P. R. China. Tobacco roots with egg masses were dipped in water to remove adhering soil, cut into 0.5 to 1 cm pieces and agitated in 0.5% sodium hypochlorite solution for 2 to 3 min. Egg masses were rinsed with sterile water then placed in 0.5% sodium hypochlorite for 2 to 3 min, agitated and rinsed with sterile water (Kerry and Bourne, 2002). The eggs were incubated at 25 ± 1°C for three to five days to obtain second stage juveniles (J2) with modified Bearmann funnel method (Gray, 1984). The J2 were then used for *in vitro* and pot experiments.

Panagrellus Redivivus

Nematodes were cultured on oatmeal medium (oatmeal: 20 g, water 80 ml) at 25°C for seven days, and then refrigerated at 4°C prior to use.

Bursaphelenchus Xylophilus

Botrytis cinerea was cultured on a potato dextrose agar (PDA) plate at 25°C until the fungus was fully grown. The plate was then inoculated with the pine nematode, and then cultured until the fungal mycelia had been completely consumed.

Caenorhabiditis Elegans

The nematodes were mutiplied in oatmeal medium at room temperature (21 to 26°C) for 6 to 7days.

In Vitro Experiment

Washed nematodes (80 to 100) were transferred to the PDA culture plates containing abundant well-developed hyphae of *Stropharia* sp. 1.2052. The plates were incubated at 25°C, counting the numbers of active and inactive nematodes under a stereomicroscope (50×) at different times (24, 48, 72, 96, 120 and 144 h). The nematodes were considered dead if they showed no response to physical stimuli. Uninoculated plates were used as negative control, and avermectin (0.8 µg/ml) was used as positive control. Each treatment was replicated three times, and the mortalities (%) were calculated.

Pot Assay

One-month-old tomato seedlings were transplanted into 15 cm-diameter pots containing 1000 g sterilized soil (loam soil : humus soil : sand, 1:1:1,v/v). A week after transplanting, the soil around the seedlings was inoculated with *M. incognita* at 2000 juveniles per plant. Pots were arranged in a randomized block in a greenhouse at 25 ± 2°C.

Four treatments were applied: three concentrations of strain 1.2052 (50, 75 and 100 g) were incorporated with soil around the seedlings. Avermectin (0.5%, 0.02 g/pot) was used as positive control, and untreated pots were used as negative control. The experiment was completely randomized with three replicates. All pots were kept under greenhouse conditions (25 to 32°C) and watered when needed. After 30 and 60 days, the plants were harvested. Nematode damage was measured by rating root galls on the 0 to 5 scale (0 = complete and healthy root system, no infection, 1 = very few small galls only be detected upon close examination, 2 = 25% of root system severely galled, 3 = 25 to 50% of root system severely galled and not, 4 = 50 to 75% of root system severely galled, 5 = >75% of root system severely galled, nearly no healthy roots). In addition, population of nematodes in soil, plant heights, and shoot fresh weights were determined.

Statistical Analysis

Results were tested with SPSS version 17.0 by one-way analysis of variance (ANOVA). Significantly differences between treatment means were separated by using *t* test at the $P<0.05$ (Sahebani and Hadavi, 2008).

RESULTS

Effect of Fungi on Nematode *in Vitro*

Application of *Stropharia* sp.1.2052 significantly inhibited nematodes motility *in vitro*, when compared to control in water. At 24 h complete inhibition was observed for all the J2 of *M. incognita* and for *P. redivivus*, whereas lower values were found for *B. xylophilus* and *C. elegans*, which required a longer exposure to reach the same level of inhibition (Table 1).

Table 1: The inhibition of *Stropharia* sp. 1.2052 on nematodes tested *in vitro*

Test time (h)	M. incognta (%)	P. redivivus (%)	C. elegans (%)	B. xylophilus (%)	Avermectin (%)	Negative control (%)
4	46.97 ± 2.96•	25.15 ± 7.46*	8.33 ± 1.43	26.01 ± 12.13	82.28 ± 9.01**	0
12	77.47 ± 22.06*	58.68 ± 6.17**	17.64 ± 1.82	54.32 ± 2.75*	91.87 ± 14.08**	2.71 ± 0.85
24	100 ± 3.45	100 ± 5.67	41.81 ± 1.88**	99.25 ± 1.29..	93.03 ± 12.06**	6.04 ± 1.24
36	100 ± 2.78	100 ± 6.77	96.81 ± 3.13**	100 ± 2.33	93.54 ± 11.19**	9.37 ± 1.89

Means of three replicates ± SD. Values in each column followed by * one and ** two asterisks were significantly different from water control at P≤0.05 and P≤0.01, respectively.

Effect of Fungi on *M. Incognita* in Pot Assay

Significant control of root-knot nematodes was observed in pots. A dose response relationship was observed for treatments with *Stropharia* isolate 1.2052 which caused the greatest decrease of galls in one and two months, compared to positive controls. Application of *Stropharia* at lower doses resulted in lower damage reduction in one and two months, respectively (Table 2).

Table 2: The control efficacy of strain 1.2052 against the *M. incognta* in pot assay

Treatment		Dosage (g)			30 days			60 days	
		Gall index	Biocontrol efficacy(%)	J2/g of plant root	J2 in soil (% in control)	Gall index	Biocontrol efficacy(%)	J2/g of root	J2 in soil (% in control)
Negative control		2	0	29.33 ± 3.45	0	2	0	86.96 ± 6.76	0
Avermectin	0.02	0	90.2	2.33 ± 0.15	98.63	0	96.18	1.33 ± 0.45	93.01
Treat1	50	1	68.2	7.67 ± 3.78*	26.39	1	45.28	24.58 ± 2.13	61.18
Treat2	75	0	84.2	4.00 ± 2.60*	41.72	0	88.14	17.03 ± 1.33	56.77
Treat3	100	0	70.6	5.35 ± 2.34	27.88	0	67.32	20.48 ± 4.53	45.77

Means of three replicates ± SD. Values in each column followed by * one and ** two asterisks were significantly different from water control at $P \leq 0.05$ and $P \leq 0.01$, respectvely.

There were significant reductions of the *M. incognita* J2 (Table 2), however a dose-response relationship was less evident as the treatment with 75 g/pot caused the highest reduction of J2 from roots, either in one and two months, compared to negative control. Application of *Stropharia* sp. 1.2052 at 50 and 100 g had 8 and 5/g roots in one month, 25 and 21/g roots in two months, respectively. The percent reduction of nematodes in soil compared to the negative control appeared dose-dependent, and was higher than avermectin.

Differences among treatment were observed for plant height and biomass. The plants showed an increase in height after two months, with tallest plants recorded for treatment with *Stropharia* sp. 1.2052 at 75 g/pot and the shortest from the positive control. The fungus application significantly improved shoot weight over the inoculated treatment, the highest weight recorded for treatment at 75 g/pot. A dose-response relationship was also observed for *Stropharia* sp. 1.2052 applied at 50, 75 and 100 g/pot that increased the fresh root weights, compared to positive and negative controls, respectively (Table 3).

Table 3: Plant growth promotion of strain 1.2052 on tomato

Treatment	Dosage (g)	Height increasing rate (%)	Fresh shoot weight (g)	Fresh root weight (g)
Negative control		78.9	12.78 ± 0.13	5.38 ± 0.34
Avermectin	0.02	74.0	8.66 ± 0.87	4.00 ± 0.29
Treat1	50	84.4	26.21 ± 1.07	10.13 ± 1.89
Treat2	75	78.8	34.25 ± 1.67*	12.68 ± 2.89
Treat3	100	79.6	32.67 ± 0.96*	19.56 ± 1.13*

Means of three replicates ± SD. Values in each column followed by * one and ** two asterisks were significantly different from water control at P≤0.05 and P≤0.01, respectvely.

DISCUSSION

Many mushrooms, previously considered only as saprobes, are actually considered capable of taking advantage of other organisms present in their microcosm, exploited for nutritional purposes (including nematodes, plants, other fungi, and bacteria). There is a special group of nematophagous basidiomycetous fungi provided with hyphal appendages that are indispensable for their nematode-attacking ability. The appendages within the various species include hourglass-shaped adhesive knobs on members of the genus *Nematoctonus* (Drechsler, 1949, 1954), and its teleomorph *Hohenbuehelia* (Barron and Dierkes, 1977). Other appendages include the secretory cells of *Pleurotus* spp. (Barron and Thorn, 1987), the secretory appendages of *Conocybelactea* (Hutchison et al., 1996), the stephanocysts on *Hyphoderma* spp. (Burdsall, 1969; Tzean and Liou, 1993) and the spiny balls produced by *Coprinus comatus* (Luo et al., 2004). Thapa et al. (1987) reported that *Stropharia rugoso-annulata* was resistant to the mycophagous nematode species *Aphelenchoides sacchari* and *Ditylenchus myceliophagus*. However, Grewal (1990) found that the mycophagous nematode *Aphelenchoides composticola* was able to feed on *S. rugoso-annulata*. *Stropharia* species also have been reported to be able to immobilize the nematode *P. redivivus* and *B. xylophilus* with spiny acanthocytes and the results shown that the acanthocytes are functional in soil (Luo et al., 2006).

In the present study, a series of experiments were undertaken to further investigate the effect of *Stropharia* sp. 1.2052 against nematodes. In the *in-vitro* assay, this isolate inhibited more than 80% of motile stages of *M. incognita*, *P. redivivus* and *B. xylophilus*, except *C. elegans*. Application of *Stropharia* sp. 1.2052 to infested tomato plants in the greenhouse reduced the number of galls and final nematode population of *M. incognita* with highest efficacy either at one and two months. Such results are similar to the nematode control effects reported for many other fungi. In controlled experiments on tomato, a pre-planting soil treatment with *P. lilacinus* strain 251 was reported to reduce root galling by 66% (Kiewnick et al., 2006). When *F. oxysporum* was inoculated on soybean, the mean number of nematodes that penetrated roots decreased by 75%, when the soil was treated with *S. rolfsii* the number decreased by 78% (Akinsanmi et al., 2003). Greenhouse study shows that *M. albus* caused significant reduction (91 and 100%) of *Paratrichodorus allius* in soil, and 100% for *Pratylenchus penetrans*in bean roots and soil; 85 and 95% for *Meloidogyne chitwoodi* in potato roots; 100% for *Meloidogyne hapla* both in pepper roots and soil (Riga et al., 2008). The trapping fungus *A. dactyloides* was applied with cow dung manure, which reduced the number of root knots by 61.7 to 66.6% and of juveniles by 68.1 to 88.0% (Kumar et al., 2006).

CONCLUSIONS

Stropharia sp. 1.2052 provided effective control of nematode comparable to those reported for other fungal species. As *Stropharia* spp. are common saprophytic basidiomycete in soil, they have less chances to be pathogenic to plants and human beings, compared to other fungi. The main problems in application of *Stropharia* are related to its cultivation and form of delivering into soil. Its use as symbiont of seeds or plant roots could also be considered for future application, as well as its development in the integrated management of root-knot nematode.

ACKNOWLEGEMENTS

This work was jointly financed by National Natural Science Foundation Program of P. R. China (30860009 and 31260446) and Natural Science Foundation of Yunnan Province (2012YN06).

REFERENCES

1. Akinsanmi OA, Adekunle OK (2003). Effect of *Fusarium oxysporum* f. sp. *glycines* and *Sclerotium rolfsii* on the pathogenicity of *Meloidogyne incognita* race 2 to soybean. Plant Soil. 253:429-435.
2. Barron GL, Dierkes Y (1977). Nematophagous fungi: *Hohenbuehelia* the perfect state of *Nematoctonus*. Can. J. Bot. 55:3054-3062.
3. Barron GL, Thorn RG (1987). Destruction of nematodes by species of *Pleurotus*. Can.J.Bot. 65:774-778.

4. Burdsall HH (1969).Stephanocysts: unique structures in the Basidiomycetes. Mycologia. 61:915-923.
5. Drechsler C (1949). A nematode-capturing fungus with anastomosing clampbearing hyphae. Mycologia 41:369-387.
6. Drechsler C (1954). A nematode-capturing fungus with clamp-connectionsand curved conidia. J. Wash Acad. Sci. 44:82-85.
7. Farr DF (1980). The acanthocyte, a unique cell type in Stropharia Agaricales. Mycotaxon. 11:241-249.
8. Gray NF (1984).Ecology of nematophagous fungi: Comparison of the soil sprinkling method with the Baerman funnel technique in the isolation of endoparasites. Soil Biol. Biochem.16:81-83.
9. Grewal PS (1990). Reproduction of the nematode Aphelenchoides composticola on cultivated mushrooms and common weed moulds. Revue de Nématologie 13:117-119.
10. Hutchison LJ, Madzia SE, Barron GL (1996). The presence and antifeedent function of toxin-producing secretory cells on hyphae of the lawn-inhabiting agaric Conocybelactea. Can. J. Bot.74:431-434.
11. Kerry BR, Bourne JM (2002). A manual for research on Verticillium chlamydosporium, a potential biological control agent for root-knot nematodes. Druckform GmbHMerckstr, Germany. p .171.
12. Kiewnick S, Sikora RA (2006). Biological control of the root-knot nematode Meloidogyne incognita by Paecilomyces lilacinus strain 251. Biol. Control. 38:179-187.
13. Koenning SR, Overstreet C, Noling JW, Donald PA, Becker JO, Fortnum BA (1999). Survey of crop losses in response to phytoparasitic nematodes in the United States for 1994. J. Nematol. 31:587-618.
14. Kumar D, Singh KP (2006). Assessment of predacity and efficacy of Arthrobotrys dactyloides for biological control of root knot disease of tomato. J. Phytopathol. 154:1-5.
15. Li MZ, Yang E, Xiang MC, Liu XZ, Chen SY (2008). Population dynamics and biocontrol efficacy of the nematophagous fungus Hirsutella rhossiliensis as affected by stage of the soybean cyst nematode. Biol. Control. 47:244-249.
16. Luo H, Li X, Li GH, Pan YB, Zhang KQ (2006). Acanthocytes of Stropharia rugosoannulata function as a nematode-attacking device. Appl. Environ. Microbiol. 72:2982-2987.

17. Luo H, Mo MH, Huang XW, Li X, Zhang KQ (2004). *Coprinus comatus*: a basidiomycete fungus forms novel spiny structures and infectsnematodes. Mycologia 96:1218-1225.

18. Manzanilla-Lo´pez RH, Eateves I, Finetti-Sialer MM, Hirsch PR (2013). *Pochonia chlamydosporia*: Advances and challenges to improve its performance as a biological control agent of sedentary endo-parasitic nematodes. J. Nematol. 45:1-7.

19. Norvell LL, Redhead SA (2000). *Stropharia albivelata* and its basionym *Pholiota albivelata*. Mycotaxon. 76:315-320.

20. Perry, RN, Moens M, Starr FJ, Eds. (2009). Root-Knot Nematodes. CAB International, Wallingford, UK.

21. Riga E, Lacey LA, Guerra N (2008). *Muscodoralbus* a potential biocontrol agent against plant-parasitic nematodes of economically important vegetable crops in Washington State. Biol. Control. 45:380-385.

22. Sahebani N, Hadavi N (2008). Biological control of the root-knot nematode *Meloidogyne javanica* by *Trichoderma harzianum*. Soil Biol. Biochem. 40:2016-2020.

23. Thapa CD, Sharma NK. (1987). Effect of temperature on pathognicity and multiplication of two myceliophagous nematodes. Indian Mushroom Sci. 2:169-172.

24. Tzean SS, Liou JY (1993). Nematophagous resupinate basidiomycetous fungi. Phytopathology 83:1015-1020.

Citations

CHAPTER 1

E. I. Koshel, V. V. Aleshin, G. A. Eroshenko1, and V, V. Kutyrev, Phylogenetic Analysis of Entomoparasitic Nematodes, Potential Control Agents of Flea Populations in Natural Foci of Plague, http://dx.doi.org/10.1155/2014/135218.

CHAPTER 2

Marie-Stéphane Tranier, Johan Pognant-Gros, Reynaldo De la Cruz Quiroz Cristóbal Noé Aguilar González, Thierry Mateille, and Sevastianos Roussos, Commercial Biological Control Agents Targeted Against Plant-Parasitic Root-knot Nematodes, doi: 10.1590/S1516-8913201402540.

CHAPTER 3

Nicolás Marro, Paola Lax, Marta Cabello, Marcelo Edmundo Doucet, and Alejandra Gabriela Becerra, Use of the Arbuscular Mycorrhizal Fungus Glomus Intraradices as Biological Control Agent of the Nematode Nacobbus Aberrans Parasitizing Tomato, doi.org/10.1590/S1516-8913201402200.

CHAPTER 4

Guy R. Knudsen and Louise-Marie C. Dandurand, "Ecological Complexity, and the Success of Fungal Biological Control Agents," Advances in Agriculture, vol. 2014, Article ID 542703, 11 pages, 2014. doi:10.1155/2014/542703.

CHAPTER 5

Sardul Singh Sandhu, Anil K. Sharma, Vikas Beniwal, Gunjan Goel, Priya Batra, Anil Kumar, Sundeep Jaglan, A. K. Sharma, and Sonal Malhotra, Myco-Biocontrol of Insect Pests: Factors Involved, Mechanism, and Regulation, doi: 10.1155/2012/126819.

CHAPTER 6

María Jesús Navarro and Francisco José Gea, Entomopathogenic nematodes for the control of phorid and sciarid flies in mushroom crops, http://dx.doi.org/10.1590/S0100-204X2014000100002

CHAPTER 7

Víctor M. Hernández-Velázquez, Laura P. Lina-García, Verónica Obregón-Barboza, Adriana G. Trejo-Loyo, and Guadalupe Peña-Chora, Pathogens Associated with Sugarcane Borers, Diatraea spp. (Lepidoptera: Crambidae): A Review, doi:10.1155/2012/303589.

CHAPTER 8

Girón-Pablo S., Ruiz-Vega J., Pérez-Pacheco R., Sánchez-García J. A., and Aquino-Bolaños T., Isolation of entomopathogenic nematodes and control of Phyllophaga vetula Horn in Oaxaca, Mexico, DOI: 10.5897/AJB12.448.

CHAPTER 9

Kevin C Kobylinski, Massamba Sylla, William Black IV, and Brian D Foy, Mermithid Nematodes Found In Adult Anopheles Southeastern Senegal, doi:10.1186/1756-3305-5-131.

CHAPTER 10

Chen Guan Pau, Chan Teck Stephen Leong, Sing King Wong, Lily Eng, Make Jiwan, Franklin Ragai Kundat, Zakry Fitri Bin Ab. Aziz, Osumanu Haruna Ahmed And Nik Muhammad Majid, Isolation of Indigenous Strains of Paecilomyces lilacinus with Antagonistic Activity against Meloidogyne incognita, ISSN: 1814–9596.

CHAPTER 11

Chuixu Kong, Chongyan Zhai, Jing Li, Jun Zhang, Keqin Zhang, and Yajun Liu, Evaluation of Stropharia sp. 1.2052 nematicidal effects against Meloidogyne incognita on tomato, DOI: 10.5897/AJMR2013.6206.

Index

N

O

P

R

S